S.W. Green's Son

Specimens of Type in Use

S.W. Green's Son

Specimens of Type in Use

ISBN/EAN: 9783337418328

Printed in Europe, USA, Canada, Australia, Japan

Cover: Foto ©berggeist007 / pixelio.de

More available books at **www.hansebooks.com**

SPECIMENS OF

IN USE BY

W. G

CO

NS FOR "CASTING   NEEDED
A GIVEN SIZED   QUAN-
ITIES OF PAPER
OF SIGNATUR
WILL
HO

NEW YORK:

S. W. GREEN'S SON, ELECTROTYPER, PRINTER AND BINDER,

74 & 76 BEEKMAN STREET.

Please do not mutilate this book.

Any of the sample pages, printed on one side only, will be furnished separately on application to

S. W. GREEN'S SON,

74 & 76 Beekman Street,

NEW YORK.

CONTENTS.

1

HOW TO "CAST-OFF," OR ESTIMATE.

AN author or publisher having in his hand a manuscript, collection of printed copy, or a combination of both, wishes to learn what shape and size of page will best suit his purpose. The first question is, How many words are there in the copy to be "put in hand"? Occasionally, calculations are made by letters instead of by words. Of course this is theoretically more accurate. A "Johnsonian" writer will not make as many words from a given number of letters as will another who uses short and simple words. But practically the words of a given composition furnish an approximately accurate measure. How shall their number be ascertained?

Take average pages and count the number of lines in each. Count in different places the words in ten consecutive lines, divide by ten, and thus get a general average of words in a line. This multiplied by the average number of lines in a page gives the average number of words in a page, and this multiplied by the number of pages gives the total number of words. With printed copy all of the same type and measure the result thus obtained ought to closely approximate correctness. So of an occasional mass of manuscript written uniformly in the same hand on the same size and ruling of paper. But generally judgment must be exercised, the copy averaged in groups most nearly alike in contents, and the results added. Suppose we have 1146 pages manuscript, 27 lines to a page, and 9.5 words to a line. The total number of words will be 293,949. This is the dividend.

Try Small Pica Old Style No. 1 leaded, 8vo, as on page 18. Divide 293,949 by 363, the number of words in that page, and we learn that the manuscript will make 810 printed pages. But so many pages are not wanted: we would like say 640. 293,949 ÷ 640 = 460 as the number of words wanted in a page. Turn to third column of Table on page 5, and Bourgeois Old Style No. 3 is the largest type leaded that will take the matter into 640 pages, and that page takes in more words (523) than are wanted. How large must a page of Long Primer No. 13 leaded (page 22) be made, to take in 460 words? Divide

460 by 16.5, the number of words to the square inch in that type, and we have 28 as the number of square inches needed. This, a page 4 × 7 inches will contain.

Or suppose the same number of words to be put into a 12mo the usual size, about 20 square inches in page. Brevier Old Style No. 3 (pages 36, 37) is the type selected. The leaded page will contain (20 × 23.1 =) 462 words, and the manuscript will make (293,949 ÷ 462 =) 637 pages. The solid page will contain (20 × 29.5 =) 590 words, and the whole will make (293,949 + 590 =) 499 pages.

These illustrations might be endlessly varied, but are enough to indicate the principle of the Table on page 5, and to assist the intelligent user to a sufficiently correct result. The page decided upon, its measurement in ems can be easily learned by the use of the type-gauge furnished with this pamphlet.

PAPER.

Next comes the question of paper. No arbitrary direction can be given as to the margin needed for any given page. Let us go back to our illustration : 640 pages 4 × 7 = 28 inches. Call the leaf 5¾ × 9. Turn to Table on page 6, and we see that 24 × 38 will print 16 pages at once. What thickness or weight shall the paper be? The thickness of "Harper's Weekly" is selected, and is known to be say 33 × 46 100 lb. Turn to page 7, take line commencing 33 × 46 and look along the line to 100. In that column, against 24 × 38, is 60, the weight needed for 24 × 38 of equal thickness with the 33 × 46 100. How much paper will be needed to print 2000? Turn to page 8 and against 640 under 16 pages to form in either outside column is 40, the number of forms. In the Table on the opposite page 9, in column 2000 against 40, is 88 Re., the amount of paper needed to print the 2000, which, as the book is to be a bound one, must be increased to 90 Re. to include the blank leaves technically called "binder's waste."

Take the other illustration, that of 637 pages 12mo, 20 square inches. Make the leaf 4¾ × 7½ inches. This needs either 23 × 41 to print 24 pages, or 30½ × 41 to print 32 pages. Perhaps the 30½ × 41 is too wide for the presses you wish to use, and you select 23 × 41. You want to make a low priced book and are satisfied with the thickness of 24 × 38 35. On page 7, over 35 in the 24 × 38 line you find 36 in the 23 × 41 line. 5000 copies are wanted. On page 8, against 648 in the 24-page column you find 27, as the number of forms, and on the opposite page under 5000 and against 27, is 148.10 Re., the amount of paper needed. Or, the presses at your command will print 30½ × 41, on which you can print 32 pages at once. Perhaps you had better make a higher priced book and use paper equal to 24 × 38 50. Page 7 shows the equivalent weight of 24 × 38 50 to be 30½ × 41 68 lb. On page 8, under 32 pages you find that 638 pages will make 20 forms, and the opposite page shows that 5000 of 20 forms needs 110 Reams.

Or, suppose you have on hand, or can most easily get, a certain size of paper: the Table below will show what sizes of leaf are at your disposal.

Hoping that these Tables and Directions, and the specimens of different types, each with its descriptive card, will greatly shorten and lighten the labor of "casting-off" for those who receive and use this pamphlet; and hoping of course that it will be the means of adding hundreds of millions of ems to the yearly product of my composing-rooms, of bringing thousands of reams of paper to my presses and tens of thousands of volumes to my bindery, and again pledging myself to give ample money's worth in all departments, believe me

The Publishing Public's most Obedient Servant,

S. W. GREEN'S SON.

Table of Sizes to which Paper will fold and trim.

Size of sheet.	8 pages to form.	16 pages to form.	24 pages to form.	32 pages to form.	36 pages to form.	64 pages to form.
inches.	inches.	inches.	inches.	inches.	inches.	inches.
19 × 24	5¾ × 9½	4½ × 5⅝	2¾ × 6	2¾ × 4½	3 × 3⅝	2½ × 2⅝
21 × 27	6½ × 10½	5 × 6½	3½ × 6½	3¼ × 4½	3¼ × 5¼	2⅞ × 3
22 × 28	6¾ × 10⅞	5¼ × 6⅞	3¼ × 7	3¼ × 5½	3½ × 4¼	2½ × 3½
23 × 33	8¼ × 10¼	5½ × 7⅝	3⅞ × 7½	3¼ × 5½	3½ × 6	2⅝ × 3¾
23 × 35	8½ × 11	5½ × 8⅞	4⅜ × 7½	4⅜ × 5¾	3½ × 5¼	2⅝ × 4
23 × 37	9 × 11½	5½ × 8½	4⅜ × 7½	4½ × 5⅝	3½ × 3½	2⅜ × 4¼
23 × 39	9½ × 11½	5½ × 9⅞	4⅞ × 7½	4⅜ × 5⅞	3½ × 6½	2⅝ × 4½
23 × 41	10 × 11½	5½ × 9½	4¾ × 7½	4⅞ × 5⅞	3½ × 6½	2⅝ × 4¾
23 × 42	10¼ × 11½	5½ × 10½	5 × 7½	5 × 5¾	3½ × 6¼	2⅝ × 4⅞
23 × 43	10½ × 11½	5½ × 10½	5½ × 7½	5½ × 5½	3½ × 6⅞	2⅝ × 5
24 × 38	9¼ × 11¾	5¼ × 9½	4½ × 7⅞	4½ × 5¾	3½ × 6	2¼ × 4⅞
24 × 40	9¾ × 11⅞	5¾ × 9⅞	4¾ × 7½	4¾ × 5⅝	3¾ × 6½	2¾ × 4⅞
24 × 42	10½ × 11⅞	5½ × 10½	5 × 7⅝	5 × 5⅞	3¾ × 6⅞	2⅝ × 5
24 × 43	10½ × 11⅞	5¾ × 10½	5½ × 7⅞	5½ × 5⅞	3¾ × 6¼	2⅜ × 5
25 × 39	9½ × 12½	6 × 9⅞	4¾ × 8	4⅝ × 5⅝	4 × 6½	2½ × 5½
26 × 38	9½ × 12½	6¼ × 9½	4½ × 8½	4½ × 6½	4½ × 6	3 × 4⅞
26 × 40	9½ × 12½	6¼ × 9½	4¾ × 8¼	4½ × 6½	4½ × 6½	3 × 4⅞
27 × 41	10 × 13½	6½ × 9½	4¾ × 8⅞	4⅞ × 6½	4¼ × 6½	3½ × 4¾
28 × 42	10¼ × 13⅞	6½ × 10½	5 × 9	5 × 6⅝	4½ × 6⅞	3¼ × 4⅞
29 × 43	10½ × 14½	7 × 10⅞	5½ × 9½	5½ × 6¾	4¾ × 6¾	3⅞ × 5
29 × 44	10½ × 14½	7 × 10⅞	5¼ × 9½	5¼ × 6⅞	4¾ × 7	3½ × 5½
30 × 42	10½ × 14⅞	7½ × 10⅝	5 × 9⅞	5 × 7½	4¾ × 6⅞	3½ × 4⅞
30 × 43	10½ × 14⅞	7½ × 10⅞	5½ × 9½	5½ × 7½	4½ × 6½	3½ × 5
30 × 44	10¾ × 14⅞	7¼ × 10½	5¾ × 9⅞	5¼ × 7½	4¾ × 7	3½ × 5½
32 × 45	11 × 15½	7¾ × 10½	5½ × 10½	5½ × 7⅞	4¾ × 7⅛	3¼ × 5¼
33 × 45	11 × 16½	8 × 10½	5⅞ × 10½	5⅞ × 7½	5¼ × 7½	3⅜ × 5¼
33 × 46	11¼ × 16½	8 × 11½	5½ × 10⅞	5½ × 7⅞	5¼ × 7½	3⅞ × 5⅞

COPYRIGHT, 1881, BY CHAS. N. GREEN.

4

Capacities of Types shown in the following pages.

Types.	Page or column in inches.	Lines in page or col.	Words in page or col.	Words in line.	Words to sq.in.	Meas-urem'nt in ems.	Words to 1000 ems.
Pica No. 6 leaded..................	25.26	32	296	9.2	11.7	943	313
" " solid....................	25.26	38	349	9.2	13.8	943	371
" Old Style No. 1 leaded.........	25.26	32	295	9.2	11.7	943	313
" " " solid..........	25.26	38	346	9.1	13.7	943	367
Small Pica No. 12 leaded...........	25.26	37	358	9.7	14.2	1269	282
" " solid.............	25.26	44	423	9.6	16.8	1269	333
" No. 9 leaded.............	25.26	37	384	10.4	15.2	1269	303
" " solid..............	25.26	44	463	10.5	18.3	1269	365
" Old Style No. 1 leaded....	25.26	37	363	9.8	14.4	1269	286
" " " solid.....	25.26	44	434	9.9	17.2	1269	342
" " No. 2 leaded....	25.26	37	373	10	14.8	1269	294
" " " solid.....	25.26	44	451	10.3	17.9	1269	356
Long Primer No. 13 leaded.........	25.26	39	418	10.7	16.5	1450	288
" " solid............	25.26	47	508	10.8	20	1450	350
" No. 9 leaded..........	25.26	39	447	11.5	17.7	1450	308
" " solid............	25.26	47	544	11.6	21.6	1450	375
" Old Style No. 2 leaded..	25.26	39	444	11.4	17.6	1450	306
" " " solid...	25.26	47	535	11.4	21.2	1450	369
" " " No. 3 leaded..	25.26	39	408	10.5	16.2	1450	281
" " " solid...	25.26	47	494	10.5	19.6	1450	340
Bourgeois No. 12 leaded...........	25.26	44	538	12¼	21.3	1824	295
" " solid..............	25.26	54	658	12.2	26	1824	361
" Old Style No. 3 leaded....	25.26	44	523	11.9	20.7	1824	287
" " " solid......	25.26	54	644	12	25.5	1824	353
Brevier No. 12 leaded..............	25.26	46	583	12.7	23.1	2135	273
" " solid................	25.26	58	739	12.7	29.3	2135	346
" Old Style No. 3 leaded......	25.26	46	584	12.7	23.1	2135	273
" " " solid.......	25.26	58	743	12.8	29.5	2135	349
Minion No. 20 leaded..............	25.26	51	695	13.6	27.5	2760	252
" " solid	25.26	66	900	13.6	35.6	2760	326
" Old Style No. 3 leaded.....	25.26	51	762	15	30.1	2760	276
" " " solid.......	25.26	66	986	14.9	39	2760	357
Nonpareil No. 20 leaded.............	25.26	58	852	14.7	33.7	3680	232
" " solid.............	25.26	77	1131	14.7	45	3680	308
" Old Style No. 3 leaded.......	25.26	58	887	15.3	35.1	3680	241
" " " solid.. ...	25.26	77	1185	15.4	47	3680	322
Agate No. 20 leaded.................	13	64	504	7.8	39	2511	200
" " solid...................	13	89	723	8.1	55	2511	288
Pearl No. 5 leaded	13	67	575	8.6	44	2842	202
" " solid...................	13	95	828	8.7	64	2842	291

A Foot will contain, of

Size.	Solid.	Leaded 6 to Pica.	Leaded 4 to Pica.	Leaded 3 to Pica.	Leaded 2 to Pica.
	lines.	lines.	lines.	lines.	lines.
Pearl....	180	146	110	98	80
Agate	168	120	105	95	77
Nonpareil..................	144	108	96	88	72
Minion....................	124	96	87	79	67
Brevier....................	109	88	79	73	62
Bourgeois..................	100	83	74	68	60
Long Primer...............	90	75	68	63	55
Small Pica.................	84	70	65	61	53
Pica	72	62	58	54	48

Sizes of Paper needed for a given sized Leaf.

Size of leaf wanted.	8 pages to form.	16 pages to form.	24 pages to form.	32 pages to form.	36 pages to form.	64 pages to form
	inches.	inches.	inches.	inches.	inches.	in.
3 × 4	8½ × 12½	12½ × 17	12¾ × 25	17½ × 25	18¾ × 25½	25 × 34
3 × 4¼	9 × 12½	12½ × 18	13½ × 25	18 × 25	18¾ × 27	25 × 36
3 × 4½	9½ × 12½	12½ × 19	14¼ × 25	18½ × 25	18¾ × 28½	25 × 37
3 × 4¾	10 × 12½	12½ × 20	15 × 25	19 × 25	18¾ × 30	25 × 38
3 × 5	10½ × 12½	12½ × 21	15¾ × 25	19½ × 25	18¾ × 31½	25 × 39
3¼ × 4¼	9 × 13½	13½ × 18	13½ × 27	18 × 27	20¼ × 27	27 × 36
3¼ × 4½	9½ × 13½	13½ × 19	14¼ × 27	19 × 27	20¼ × 28½	27 × 38
3¼ × 4¾	10 × 13½	13½ × 20	15 × 27	20 × 27	20¼ × 30	27 × 40
3¼ × 5	10½ × 13½	13½ × 21	15¾ × 27	21 × 27	20¼ × 31½	27 × 42
3¼ × 5¼	11 × 13½	13½ × 22	16½ × 27	22 × 27	20½ × 33	27 × 44
3½ × 4½	9¾ × 15	15 × 19½	14⅝ × 30	19½ × 30	22½ × 29¼	30 × 39
3½ × 4¾	10¼ × 15	15 × 20½	15⅜ × 30	20½ × 30	22½ × 30¾	30 × 41
3½ × 5	10¾ × 15	15 × 21½	16⅛ × 30	21½ × 30	22½ × 32¼	30 × 43
3½ × 5¼	11¼ × 15	15 × 22½	16⅞ × 30	22½ × 30	22½ × 33¾	30 × 45
3½ × 5½	11¾ × 15	15 × 23½	17⅝ × 30	23½ × 30	22½ × 35¼	....
3¾ × 4¾	10¼ × 16	16 × 20½	15⅜ × 32	20½ × 32	24 × 30¾	32 × 41
3¾ × 5	10¾ × 16	16 × 21½	16⅛ × 32	21½ × 32	24 × 32¼	32 × 43
3⅞ × 5¼	11¼ × 16	16 × 22½	16⅞ × 32	22½ × 32	24 × 33¾	32 × 45
3⅞ × 5½	11¾ × 16	16 × 23½	17⅝ × 32	23½ × 32	24 × 35¼	....
3⅞ × 5¾	12¼ × 16	16 × 24½	18⅜ × 32	24½ × 32	24 × 36¾	....
4 × 5	10¾ × 17	17 × 21½	16⅛ × 34	21½ × 34	25½ × 32¼	....
4 × 5¼	11¼ × 17	17 × 22½	16⅞ × 34	22½ × 34	25½ × 33¾	....
4 × 5½	11¾ × 17	17 × 23½	17⅝ × 34	23½ × 34	25½ × 35¼	....
4 × 5¾	12¼ × 17	17 × 24½	18⅜ × 34	24½ × 34	25½ × 36¾	....
4 × 6	12¾ × 17	17 × 25½	19⅛ × 34	25½ × 34	25½ × 38¼	....
4¼ × 5¼	11¼ × 18	18 × 22½	16⅞ × 36	22½ × 36	27 × 33¾	....
4¼ × 5½	11¾ × 18	18 × 23½	17⅝ × 36	23½ × 36	27 × 35¼	....
4¼ × 5¾	12¼ × 18	18 × 24½	18⅜ × 36	24½ × 36	27 × 36¾	....
4¼ × 6	12¾ × 18	18 × 25½	19⅛ × 36	25½ × 36	27 × 38¼	....
4¼ × 6¼	13¼ × 18	18 × 26½	19⅞ × 36	26½ × 36	27 × 39¾	....
4¼ × 6½	13¾ × 18	18 × 27½	20⅝ × 36	27½ × 36	27 × 41¼	....
4¼ × 6¾	14¼ × 18	18 × 28½	21⅜ × 36	28½ × 36	27 × 42¾	....
4½ × 5½	11¾ × 19	19 × 23½	17⅝ × 38	23½ × 38	28½ × 35¼	....
4½ × 5¾	12¼ × 19	19 × 24½	18⅜ × 38	24½ × 38	28½ × 36¾	....
4½ × 6	12¾ × 19	19 × 25½	19⅛ × 38	25½ × 38	28½ × 38¼	....
4½ × 6¼	13¼ × 19	19 × 26½	19⅞ × 38	26½ × 38	28½ × 39¾	....
4½ × 6½	13¾ × 19	19 × 27½	20⅝ × 38	27½ × 38	28½ × 41¼	....
4½ × 6¾	14¼ × 19	19 × 28½	21⅜ × 38	28½ × 38	28½ × 42¾	....
4½ × 7	14¾ × 19	19 × 29½	22⅛ × 38	29½ × 38	28½ × 44¼	....
4¾ × 5¾	12¼ × 20	20 × 24½	18⅜ × 40	24½ × 40	30 × 36¾	....
4¾ × 6	12¾ × 20	20 × 25½	19⅛ × 40	25½ × 40	30 × 38¼	....
4¾ × 6¼	13¼ × 20	20 × 26½	19⅞ × 40	26½ × 40	30 × 39¾	....
4¾ × 6½	13¾ × 20	20 × 27½	20⅝ × 40	27½ × 40	30 × 41¼	....
4¾ × 6¾	14¼ × 20	20 × 28½	21⅜ × 40	28½ × 40	30 × 42¾	....
4¾ × 7	14¾ × 20	20 × 29½	22⅛ × 40	29½ × 40	30 × 44¼	....
4¾ × 7½	15¼ × 20	20 × 30½	23 × 41	30½ × 41	30 × 45¾	....
5 × 6¼	13¼ × 21	21 × 26½	19⅞ × 42	26½ × 42	31½ × 39¾	....
5 × 6½	13¾ × 21	21 × 27½	20⅝ × 42	27½ × 42	31½ × 41¼	....
5 × 6¾	14¼ × 21	21 × 28½	21⅜ × 42	28½ × 42	31½ × 42¾	....
5 × 7	14¾ × 21	21 × 29½	22⅛ × 42	29½ × 42	31½ × 44¼	....
5 × 7¼	15¼ × 21	21 × 30½	22⅞ × 42	30½ × 42	31½ × 45¾	....
5 × 7½	15¾ × 21	21 × 31½	23⅝ × 42	31½ × 42	31½ × 47¼	....
5¼ × 6½	13¾ × 22	22 × 27½	20⅝ × 44	27½ × 44	33 × 41¼	....
5¼ × 6¾	14¼ × 22	22 × 28½	21⅜ × 44	28½ × 44	33 × 42¾	....
5¼ × 7	14¾ × 22	22 × 29½	22⅛ × 44	29½ × 44	33 × 44¼	....
5¼ × 7¼	15¼ × 22	22 × 30½	22⅞ × 44	30½ × 44	33 × 45¾	....
5¼ × 7½	15¾ × 22	22 × 31½	23⅝ × 44	31½ × 44	....	....
5¼ × 7¾	16¼ × 22	22 × 32½	24⅜ × 44	32½ × 44	....	....
5¼ × 8	16¾ × 22	22 × 33½	25⅛ × 44	33½ × 44	....	....
5½ × 6½	13¾ × 23	23 × 27½	20⅝ × 46	27½ × 46	....	....
5½ × 6¾	14¼ × 23	23 × 28½	21⅜ × 46	28½ × 46	....	....
5½ × 7	14¾ × 23	23 × 29½	22⅛ × 46	29½ × 46	....	....
5½ × 7¼	15¼ × 23	23 × 30½	22⅞ × 46	30½ × 46	....	....
5½ × 7½	15¾ × 23	23 × 31½	23⅝ × 46	31½ × 46	....	....
5½ × 7¾	16¼ × 23	23 × 32½	24⅜ × 46	32½ × 46	....	....
5½ × 8	16¾ × 23	23 × 33½	25⅛ × 46	33½ × 46	....	....

Size of leaf wanted.	8 pages to form.	16 pages to form.
	inches.	inches.
5¾ × 6¾	14¼ × 24	24 × 28½
5¾ × 7	14¾ × 24	24 × 29½
5¾ × 7¼	15¼ × 24	24 × 30½
5¾ × 7½	15¾ × 24	24 × 31½
5¾ × 7¾	16¼ × 24	24 × 32½
5¾ × 8	16¾ × 24	24 × 33½
5¾ × 8¼	17¼ × 24	24 × 34½
5¾ × 8½	17¾ × 24	24 × 35½
5¾ × 8¾	18½ × 24	24 × 36½
5¾ × 9	18¾ × 24	24 × 37½
6 × 7	14¾ × 25	25 × 29½
6 × 7¼	15¼ × 25	25 × 30½
6 × 7½	15¾ × 25	25 × 31½
6 × 7¾	16¼ × 25	25 × 32½
6 × 8	16¾ × 25	25 × 33½
6 × 8¼	17¼ × 25	25 × 34½
6 × 8½	17¾ × 25	25 × 35½
6 × 8¾	18¼ × 25	25 × 36½
6 × 9	18¾ × 25	25 × 37½
6 × 9½	19¾ × 25	25 × 39½
6¼ × 7¼	15¼ × 26	26 × 30½
6¼ × 7½	15¾ × 26	26 × 31½
6¼ × 7¾	16¼ × 26	26 × 32½
6¼ × 8	16¾ × 26	26 × 33½
6¼ × 8¼	17¼ × 26	26 × 34½
6¼ × 8½	17¾ × 26	26 × 35½
6¼ × 8¾	18¼ × 26	26 × 36½
6¼ × 9	18¾ × 26	26 × 37½
6¼ × 9¼	19¼ × 26	26 × 38½
6¼ × 9½	19¾ × 26	26 × 39½
6¼ × 9¾	20¼ × 26	26 × 40½
6½ × 7½	15¾ × 27	27 × 31½
6½ × 7¾	16¼ × 27	27 × 32½
6½ × 8	16¾ × 27	27 × 33½
6½ × 8¼	17¼ × 27	27 × 34½
6½ × 8½	17¾ × 27	27 × 35½
6½ × 8¾	18¼ × 27	27 × 36½
6½ × 9	18¾ × 27	27 × 37½
6½ × 9¼	19¼ × 27	27 × 38½
6½ × 9½	19¾ × 27	27 × 39½
6½ × 9¾	20¼ × 27	27 × 40½
6½ × 10	20¾ × 27	27 × 41½
6½ × 10¼	21¼ × 27	27 × 42½
6¾ × 7¾	16¼ × 28	28 × 32½
6¾ × 8	16¾ × 28	28 × 33½
6¾ × 8¼	17¼ × 28	28 × 34½
6¾ × 8½	17¾ × 28	28 × 35½
6¾ × 8¾	18¼ × 28	28 × 36½
6¾ × 9	18¾ × 28	28 × 37½
6¾ × 9¼	19¼ × 28	28 × 38½
6¾ × 9½	19¾ × 28	28 × 39½
6¾ × 9¾	20¼ × 28	28 × 40½
6¾ × 10	20¾ × 28	28 × 41½
6¾ × 10¼	21¼ × 28	28 × 42½
6¾ × 10½	21¾ × 28	28 × 43½
7 × 8	16¾ × 29	29 × 33½
7 × 8¼	17¼ × 29	29 × 34½
7 × 8½	17¾ × 29	29 × 35½
7 × 8¾	18¼ × 29	29 × 36½
7 × 9	18¾ × 29	29 × 37½
7 × 9¼	19¼ × 29	29 × 38½
7 × 9½	19¾ × 29	29 × 39½
7 × 9¾	20¼ × 29	29 × 40½
7 × 10	20¾ × 29	29 × 41½
7 × 10¼	21¼ × 29	29 × 42½
7 × 10½	21¾ × 29	29 × 43½
7 × 10¾	22¼ × 29	29 × 44½
7¼ × 8¼	17¼ × 30	30 × 34½
7¼ × 8½	17¾ × 30	30 × 35½
7¼ × 8¾	18¼ × 30	30 × 36½

Size of leaf.	8 pages to form.	16 pages to form.
7¼ × 9	18¾ × 30	30 × 37½
7¼ × 9¼	19¼ × 30	30 × 38½
7¼ × 9½	19¾ × 30	30 × 39½
7¼ × 9¾	20¼ × 30	30 × 40½
7¼ × 10	20¾ × 30	30 × 41½
7¼ × 10¼	21¼ × 30	30 × 42½
7¼ × 10½	21¾ × 30	30 × 43½
7¼ × 10¾	22¼ × 30	30 × 44½
7¼ × 11	22¾ × 30	30 × 45½

Size of leaf.	8 pages to form.	16 pages to form.
8 × 9	18¾ × 33	33 × 37½
8 × 9½	19¾ × 33	33 × 39½
8 × 10	20¾ × 33	33 × 41½
8 × 10½	21¾ × 33	33 × 43½
8 × 11	22¾ × 33	33 × 45½
8 × 11½	23¾ × 33	....
8 × 12	24¾ × 33	....
8 × 12½	25¾ × 33	....
8 × 13	26¾ × 33	....

Size of leaf.	8 pages to form.
9 × 10	20¾ × 37
9 × 10½	21¾ × 37
9 × 11	22¾ × 37
9 × 11½	23¾ × 37
9 × 12	24¾ × 37
9 × 12½	25¾ × 37
9 × 13	26¾ × 37
9 × 13½	27¾ × 37
9 × 14	28¾ × 37

Size of leaf.	8 pages to form.
10 × 12	24¾ × 41
10 × 13	26¾ × 41
10 × 14	28¾ × 41
11 × 13	26¾ × 45
11 × 14	28¾ × 45
11 × 15	30¾ × 45
11 × 15⅝	32 × 45
11¼ × 16½	33 × 46

Equivalent Weights of Paper.

SIZE.	sq. inches	lbs.	lbs.	lbs.	lbs.	lbs.	lbs.	lbs.	lbs.	lbs.	lbs.	lbs.	lbs.	lbs.	lbs.	lbs.	lbs.
14 × 17	238	7	7	8	9	10	10	12	12	13	13	15	16	17	18	20	21
17 × 22	374	10	11	12	14	16	16	18	19	20	21	23	25	27	29	31	33
18 × 23	414	11	13	14	16	17	18	20	22	22	23	25	27	30	32	34	36
19 × 24	456	13	14	15	18	19	20	22	23	24	25	28	30	33	35	37	40
20 × 25	500	14	15	17	19	21	22	24	25	26	28	31	33	36	39	41	44
22 × 28	616	17	19	19	24	26	27	30	31	32	34	38	40	44	47	51	54
22 × 29	638	18	19	21	25	27	28	31	32	34	35	39	42	46	49	52	56
23 × 30	690	19	21	23	27	29	30	33	35	36	38	42	46	49	53	57	61
23 × 32	736	20	22	25	28	31	32	36	37	39	40	45	48	53	57	60	65
23 × 33	759	21	23	25	29	32	33	37	38	40	42	47	50	54	59	63	67
23 × 36	828	23	25	28	32	35	36	40	41	44	46	51	54	59	64	68	73
23 × 37	851	23	26	28	33	35	37	41	43	45	48	52	56	61	66	70	75
23 × 38	874	24	27	29	34	36	38	42	44	46	48	54	58	62	67	72	77
23 × 39	897	25	27	30	35	37	39	43	45	47	49	55	59	64	69	74	79
23 × 41	943	26	29	31	36	39	41	46	47	50	52	58	62	67	73	77	83
23 × 42	966	27	29	32	37	40	42	47	48	51	53	59	64	69	74	79	85
23 × 43	989	27	30	33	38	41	43	48	49	52	54	61	65	71	76	81	87
24 × 30	720	20	22	24	28	30	32	35	36	38	40	46	47	51	55	59	63
24 × 32	768	21	23	26	30	32	34	37	38	40	42	47	50	55	59	63	67
24 × 33	792	22	24	26	30	33	35	38	40	42	44	49	52	57	61	65	69
24 × 35	840	23	25	28	32	35	37	41	42	44	46	52	55	60	65	69	74
24 × 36	864	24	26	29	33	36	38	42	43	45	48	53	57	62	66	71	76
24 × 37	888	24	27	30	34	37	39	43	44	47	49	55	58	63	68	73	78
24 × 38	912	25	28	30	35	38	40	44	45	48	50	56	60	65	70	75	80
24 × 39	936	26	28	31	36	39	41	45	47	49	51	57	62	67	72	77	82
24 × 40	960	26	28	32	37	40	42	46	48	51	53	59	63	69	74	79	84
25 × 37	925	26	28	31	36	39	40	45	46	49	51	57	61	66	71	76	81
25 × 38	950	26	29	32	37	40	42	46	47	50	52	58	63	68	73	78	83
25 × 39	975	27	30	32	38	41	43	47	49	51	54	60	64	70	75	80	86
25 × 40	1000	27	30	33	38	42	44	48	50	53	55	60	66	72	77	82	88
25 × 41	1025	28	31	34	39	43	45	50	51	54	56	63	68	73	79	84	90
25 × 42	1050	29	32	35	40	44	46	51	53	55	58	64	69	75	81	86	92
25 × 43	1075	30	33	36	41	45	47	52	54	57	59	66	71	77	83	88	94
25 × 44	1100	30	33	37	42	46	48	53	55	58	60	63	72	79	85	90	97
26 × 38	988	27	30	33	38	41	43	48	49	50	54	63	65	71	76	81	87
26 × 39	1014	28	31	34	39	42	44	49	51	53	56	62	67	72	78	83	89
26 × 40	1040	29	32	35	40	43	46	50	52	55	57	64	68	74	80	85	91
26 × 41	1066	29	32	36	41	44	47	52	53	56	59	65	70	76	82	87	93
26 × 43	1118	30	34	37	43	46	49	54	56	59	61	69	74	80	86	92	98
27 × 39	1053	29	32	35	40	44	46	51	53	55	58	65	69	75	81	86	92
27 × 40	1080	30	33	36	42	45	47	52	54	57	59	66	71	77	83	89	95
27 × 41	1107	30	34	37	43	46	48	54	55	58	61	68	73	79	85	91	97
27 × 42	1134	31	34	38	44	47	50	55	57	60	62	70	75	81	87	93	99
27 × 44	1188	33	36	40	46	49	52	57	59	63	65	73	78	85	91	97	104
28 × 38	1064	29	32	35	41	44	47	51	53	56	59	66	70	76	82	87	93
28 × 39	1092	30	33	36	42	45	48	52	55	58	60	67	72	78	84	90	96
28 × 40	1120	31	34	37	43	47	49	54	56	59	62	69	74	80	86	92	98
28 × 41	1148	31	35	38	44	48	50	55	57	60	63	70	76	82	88	94	101
28 × 42	1176	32	36	39	45	49	52	57	59	62	65	72	77	84	90	96	103
28 × 43	1204	33	37	40	46	50	53	58	60	63	66	74	78	86	93	99	106
28 × 44	1232	34	37	41	47	51	54	60	62	65	68	76	81	88	95	101	108
29 × 40	1160	32	35	39	45	48	51	56	58	61	64	71	76	83	89	95	102
29 × 41	1189	33	36	40	46	50	52	57	60	63	65	73	78	85	91	98	104
29 × 42	1218	33	37	41	47	51	53	59	61	64	67	75	80	87	94	100	107
29 × 43	1247	34	38	42	48	52	55	60	62	66	69	78	82	89	96	102	109
29 × 44	1276	35	39	43	49	53	56	62	64	67	70	78	84	91	98	105	112
30 × 40	1200	33	36	40	46	50	53	58	60	63	66	74	79	86	93	99	105
30 × 41	1230	34	37	41	47	51	54	59	61	65	68	76	81	88	95	101	108
30 × 42	1260	35	38	42	48	52	55	61	63	66	69	77	83	90	97	103	111
30 × 43	1290	35	40	43	50	54	57	62	64	68	70	79	85	92	99	106	113
31 × 40	1240	34	38	41	48	52	54	60	62	65	69	77	82	89	96	102	109
31 × 42	1302	36	39	43	50	54	57	63	65	69	72	80	86	93	100	107	114
31 × 43	1333	37	40	44	51	56	58	64	67	70	73	82	88	95	103	109	117
31 × 44	1364	37	41	45	52	57	60	66	68	72	75	84	90	97	105	112	120
31 × 45	1395	38	42	46	54	58	61	67	70	73	77	86	92	100	107	114	122
31 × 46	1426	39	43	48	55	59	63	69	71	75	78	87	94	102	110	117	125
32 × 45	1440	39	44	48	55	60	63	70	72	76	79	88	95	103	111	118	127
32 × 46	1472	40	45	49	57	61	65	71	74	77	81	90	97	105	113	121	129
33 × 46	1518	42	46	51	58	63	67	73	76	80	83	93	100	108	117	124	133
34 × 45	1530	42	46	51	59	64	67	74	77	81	84	94	101	109	118	125	134
34 × 46	1564	43	47	52	60	65	69	76	78	82	86	96	103	112	120	128	137
34 × 47	1598	44	48	53	61	67	70	77	80	84	88	98	105	114	123	130	140
34 × 48	1632	45	49	54	63	68	72	79	82	86	90	100	107	117	126	133	143
35 × 48	1680	46	51	56	65	70	72	81	84	85	92	103	111	120	129	137	147
35 × 49	1715	47	52	57	66	71	75	83	86	90	94	105	113	123	132	140	150
35 × 50	1750	48	53	58	67	73	77	84	87	92	96	107	115	125	135	143	153
36 × 49	1764	48	53	59	68	74	77	85	88	93	97	108	116	126	136	144	154
36 × 50	1800	50	55	60	69	75	79	87	90	95	99	110	118	129	139	147	158
36 × 51	1836	51	56	61	71	76	81	89	92	97	101	113	120	131	141	150	161
36 × 52	1872	51	57	62	72	78	82	90	94	99	103	115	124	134	144	153	164

Wanted, for example, the weight of 23 × 41 paper to equal 33 × 46=100. Look along the line commencing 33 × 46 to the column in which 100 stands ; above, in that column opposite 23 × 41 is 62, which is the required weight.

TABLE OF SIGNATURES.

No. of Sig.	8 PAGES TO FORM.		12 PAGES TO FORM.		16 PAGES TO FORM.		24 PAGES TO FORM.		32 PAGES TO FORM.		36 PAGES TO FORM.		No. of Sig.
	First.	Last.	First.	Last.	First.	Last.	First.	Last.	First.	Last.	First.	Last.	
1	1	8	1	12	1	16	1	24	1	32	1	36	1
2	9	16	13	24	17	32	25	48	33	64	37	72	2
3	17	24	25	36	33	48	49	72	65	96	73	108	3
4	25	32	37	48	49	64	73	96	97	128	109	144	4
5	33	40	49	60	65	80	97	120	129	160	145	180	5
6	41	48	61	72	81	96	121	144	161	192	181	216	6
7	49	56	73	84	97	112	145	168	193	224	217	252	7
8	57	64	85	96	113	128	169	192	225	256	253	288	8
9	65	72	97	108	129	144	193	216	257	288	289	324	9
10	73	80	109	120	145	160	217	240	289	320	325	360	10
11	81	88	121	132	161	176	241	264	321	352	361	396	11
12	89	96	133	144	177	192	265	288	353	384	397	432	12
13	97	104	145	156	193	208	289	312	385	416	433	468	13
14	105	112	157	168	209	224	313	336	417	448	469	504	14
15	113	120	169	180	225	240	337	360	449	480	505	540	15
16	121	128	181	192	241	256	361	384	481	512	541	576	16
17	129	136	193	204	257	272	385	408	513	544	577	612	17
18	137	144	205	216	273	288	409	432	545	576	613	648	18
19	145	152	217	228	289	304	433	456	577	608	649	684	19
20	153	160	229	240	305	320	457	480	609	640	685	720	20
21	161	168	241	252	321	336	481	504	641	672	721	756	21
22	169	176	253	264	337	352	505	528	673	704	757	792	22
23	177	184	265	276	353	368	529	552	705	736	793	828	23
24	185	192	277	288	369	384	553	576	737	768	829	864	24
25	193	200	289	300	385	400	577	600	769	800	865	900	25
26	201	208	301	312	401	416	601	624	801	832	901	936	26
27	209	216	313	324	417	432	625	648	833	864	937	972	27
28	217	224	325	336	433	448	649	672	865	896	973	1008	28
29	225	232	337	348	449	464	673	696	897	928	1009	1044	29
30	233	240	349	360	465	480	697	720	929	960	1045	1080	30
31	241	248	361	372	481	496	721	744	961	992	1081	1116	31
32	249	256	373	384	497	512	745	768	993	1024	1117	1152	32
33	257	264	385	396	513	528	769	792	1025	1056	1153	1188	33
34	265	272	397	408	529	544	793	816	1057	1088	1189	1224	34
35	273	280	409	420	545	560	817	840	1089	1120	1225	1260	35
36	281	288	421	432	561	576	841	864	1121	1152	1261	1296	36
37	289	296	433	444	577	592	865	888	1153	1184	1297	1332	37
38	297	304	445	456	593	608	889	912	1185	1216	1333	1368	38
39	305	312	457	468	609	624	913	936	1217	1248	1369	1404	39
40	313	320	469	480	625	640	937	960	1249	1280	1405	1440	40
41	321	328	481	492	641	656	961	984	1281	1312	1441	1476	41
42	329	336	493	504	657	672	985	1008	1313	1344	1477	1512	42
43	337	344	505	516	673	688	1009	1032	1345	1376	1513	1548	43
44	345	352	517	528	689	704	1033	1056	1377	1408	1549	1584	44
45	353	360	529	540	705	720	1057	1080	1409	1440	1585	1620	45
46	361	368	541	552	721	736	1081	1104	1441	1472	1621	1656	46
47	369	376	553	564	737	752	1105	1128	1473	1504	1657	1692	47
48	377	384	565	576	753	768	1129	1152	1505	1536	1693	1728	48
49	385	392	577	588	769	784	1153	1176	1537	1568	1729	1764	49
50	393	400	589	600	785	800	1177	1200	1569	1600	1765	1800	50
51	401	408	601	612	801	816	1201	1224	1601	1632	1801	1836	51
52	409	416	613	624	817	832	1225	1248	1633	1664	1837	1872	52
53	417	424	625	636	833	848	1249	1272	1665	1696	1873	1908	53
54	425	432	637	644	849	864	1273	1296	1697	1728	1909	1944	54
55	433	440	645	660	865	880	1297	1320	1729	1760	1945	1980	55
56	441	448	661	672	881	896	1321	1344	1761	1792	1981	2016	56
57	449	456	673	684	897	912	1345	1368	1793	1824	2017	2052	57
58	457	464	685	696	913	928	1369	1392	1825	1856	2053	2088	58
59	465	472	697	708	929	944	1393	1414	1857	1888	2089	2124	59
60	473	480	709	720	945	960	1415	1440	1889	1920	2125	2160	60

QUANTITIES OF PAPER REQUIRED, 480 SHEETS TO REAM.

Sig.	250	500	750	1000	1500	2000	2500	3000	5000	10,000	Sig.
	Re. Qu.	Re. Qu.	Re. Qu.	Re. Qu.	Re. Qu.	Re. Qu.	Re. Qu.	Re. Qu.	Re. Qu.	Re. Qu.	
1	06	12	17	1 02	1 13	2 04	2 15	3 06	5 10	10 15	1
2	12	1 03	1 13	2 04	3 06	4 08	5 10	6 12	11	21 10	2
3	18	1 15	2 10	3 06	4 19	6 12	8 05	9 18	16 50	32 05	3
4	1 04	2 06	3 06	4 08	6 12	8 16	11	13 04	22	43	4
5	1 10	2 18	4 03	5 10	8 05	11	13 15	16 10	27 10	53 15	5
6	1 16	3 09	4 19	6 12	9 18	13 04	16 10	19 16	33	64 10	6
7	2 02	4 01	5 16	7 14	11 11	15 08	19 05	23 02	38 10	75 05	7
8	2 08	4 12	6 12	8 16	13 04	17 12	22	26 08	44	86	8
9	2 14	5 04	7 09	9 18	14 17	19 16	24 15	29 14	49 10	96 15	9
10	3	5 15	8 05	11	16 10	22	27 10	33	55	107 10	10
11	3 06	6 07	9 02	12 02	18 03	24 04	30 05	36 06	60 10	118 05	11
12	3 12	6 18	9 18	13 04	19 16	26 08	33	39 12	66	129	12
13	3 18	7 10	10 15	14 06	21 09	28 12	35 15	42 18	71 10	139 15	13
14	4 04	8 01	11 11	15 08	23 02	30 16	38 10	46 04	77	150 10	14
15	4 10	8 13	12 08	16 10	24 15	33	41 05	49 10	82 10	161 05	15
16	4 16	9 04	13 04	17 12	26 08	35 04	44	52 16	88	172	16
17	5 02	9 16	14 01	18 14	28 01	37 08	46 15	56 02	93 10	182 15	17
18	5 08	10 07	14 17	19 16	29 14	39 12	49 10	59 08	99	193 10	18
19	5 14	10 19	15 14	20 18	31 07	41 16	52 05	62 14	104 10	204 05	19
20	6	11 10	16 10	22	33	44	55	66	110	215	20
21	6 06	12 02	17 07	23 02	34 13	46 04	57 15	69 06	115 10	225 15	21
22	6 12	12 13	18 03	24 04	36 06	48 08	60 10	72 12	121	236 10	22
23	6 18	13 05	19	25 06	37 19	50 12	63 05	75 18	126 10	247 05	23
24	7 04	13 16	19 16	26 08	39 12	52 16	66	79 04	132	258	24
25	7 10	14 08	20 13	27 10	41 05	55	68 15	82 10	137 10	268 15	25
26	7 16	14 19	21 09	28 12	42 18	57 04	71 10	85 16	143	279 10	26
27	8 02	15 11	22 06	29 14	44 11	59 08	74 05	89 02	148 10	290 05	27
28	8 08	16 02	23 02	30 16	46 04	61 12	77	92 08	154	301	28
29	8 14	16 14	23 19	31 18	47 17	63 14	79 15	95 14	159 10	311 15	29
30	9	17 05	24 15	33	49 10	66	82 10	99	165	322 10	30
31	9 06	17 17	25 12	34 02	51 03	68 04	85 05	102 06	170 10	333 05	31
32	9 12	18 08	26 08	35 04	52 16	70 08	88	105 12	176	344	32
33	9 18	19	27 05	36 06	54 09	72 12	90 15	108 18	181 10	354 15	33
34	10 04	19 11	28 01	37 08	56 02	74 16	93 10	112 04	187	365 10	34
35	10 10	20 03	28 18	38 10	57 15	77	96 05	115 10	192 10	376 05	35
36	10 16	20 14	29 14	39 12	59 08	79 04	99	118 16	198	387	36
37	11 02	21 06	30 11	40 14	61 01	81 08	101 15	122 02	203 10	397 15	37
38	11 08	21 17	31 07	41 16	62 14	83 12	104 10	125 08	209	408 10	38
39	11 04	22 09	32 04	42 18	64 07	85 16	107 05	128 14	214 10	419 05	39
40	12	23	33	44	66	88	110	132	220	430	40
41	12 06	23 12	33 17	45 02	67 13	90 04	112 15	135 06	225 10	440 15	41
42	12 12	24 03	34 13	46 04	69 06	92 08	115 10	138 12	231	451 10	42
43	12 18	24 15	35 10	47 06	70 19	94 12	118 05	141 18	236 10	462 05	43
44	13 04	25 06	36 06	48 08	72 12	96 16	121	145 04	242	473	44
45	13 10	25 18	37 03	49 10	74 05	99	123 15	148 10	247 10	483 15	45
46	13 16	26 09	37 19	50 12	75 18	101 04	126 10	151 16	253	494 10	46
47	14 02	27 01	38 16	51 14	77 11	103 08	129 05	155 02	258 10	505 05	47
48	14 08	27 12	39 12	52 16	79 04	105 12	132	158 08	264	516	48
49	14 14	28 04	40 09	53 18	80 17	107 16	134 15	161 14	269 10	526 15	49
50	15	28 15	41 05	55	82 10	110	137 10	165	275	537 10	50
51	15 06	29 07	42 02	56 02	84 03	112 04	140 05	168 06	280 10	548 05	51
52	15 12	29 18	42 18	57 04	85 16	114 08	143	171 12	286	559	52
53	15 18	30 10	43 15	58 06	87 09	116 12	145 15	174 18	291 10	569 15	53
54	16 04	31 01	44 11	59 08	89 02	118 16	148 10	178 04	297	580 10	54
55	16 10	31 13	45 08	60 10	90 15	121	151 05	181 10	302 10	591 05	55
56	16 16	32 04	46 04	61 12	92 08	123 04	154	184 16	308	602	56
57	17 02	32 16	47 01	62 14	94 01	125 08	156 15	188 02	313 10	612 15	57
58	17 08	33 07	47 17	63 16	95 14	127 12	159 10	191 08	319	623 10	58
59	17 14	33 19	48 14	64 18	97 07	129 16	162 05	194 14	324 10	634 05	59
60	18	34 10	49 10	66	99	132	165	198	330	645	60

So I am content to tell my simple story, without trying to make things seem better than they were; dreading nothing, indeed, but falsity, which, in spite of one's best efforts, there is reason to dread. Falsehood is so easy, truth so difficult. The pencil is conscious of a delightful facility in drawing a griffin—the longer the claws, and the larger the wings, the better; but that marvellous facility, which we mistook for genius, is apt to forsake us when we want to draw a real unexaggerated lion. Examine your words well, and you will find that even when you have no motive to be false, it is a very hard thing to say the exact truth, even about your own immediate feelings—much harder than to say something fine about them which is *not* the exact truth.

It is for this rare, precious quality of truthfulness that I delight in many Dutch paintings, which lofty-minded people despise. I find a source of delicious sympathy in these faithful pictures of a monotonous homely existence, which has been the fate of so many more among my fellow-mortals than a life of pomp or of absolute indigence, of tragic suffering or of world-stirring actions. I turn without shrinking, from cloud-borne angels, from prophets, sibyls, and heroic war-riors, to an old woman bending over her flower-pot, or eating her solitary dinner, while the noonday light, softened, perhaps, by a screen of leaves, falls on her mob-cap, and just touches the rim of her spinning-wheel, and her stone jug, and all those cheap, common things which are the precious necessaries of life to her; or I turn to that village wedding, kept between four brown walls, where an awkward bridegroom opens the

10

PICA No. 6 LEADED
32 lines 296 words in page
9·2 words to line
Page 3¼ × 6¾ = 25·26 square inches
1 1·7 words to square inch
Page measures 41 × 23 = 943 ems
313 words to 1000 ems

S. W. GREEN'S SON
COMPOSITION, ELECTROTYPING, PRESSWORK
AND BINDING
74 and 76 Beekman Street
NEW YORK

So I am content to tell my simple story, without trying to make things seem better than they were; dreading nothing, indeed, but falsity, which, in spite of one's best efforts, there is reason to dread. Falsehood is so easy, truth so difficult. The pencil is conscious of a delightful facility in drawing a griffin—the longer the claws, and the larger the wings, the better; but that marvellous facility, which we mistook for genius, is apt to forsake us when we want to draw a real unexaggerated lion. Examine your words well, and you will find that even when you have no motive to be false, it is a very hard thing to say the exact truth, even about your own immediate feelings— much harder than to say something fine about them which is *not* the exact truth.

It is for this rare, precious quality of truthfulness that I delight in many Dutch paintings, which loftyminded people despise. I find a source of delicious sympathy in these faithful pictures of a monotonous homely existence, which has been the fate of so many more among my fellow-mortals than a life of pomp or of absolute indigence, of tragic suffering or of world-stirring actions. I turn without shrinking, from cloud-borne angels, from prophets, sibyls, and heroic warriors, to an old woman bending over her flower-pot, or eating her solitary dinner, while the noonday light, softened, perhaps, by a screen of leaves, falls on her mob-cap, and just touches the rim of her spinning-wheel, and her stone jug, and all those cheap, common things which are the precious necessaries of life to her; or I turn to that village wedding, kept between four brown walls, where an awkward bridegroom opens the dance with a high-shouldered, broad-faced bride, while elderly and middle-aged friends look on, with very irregular noses and lips, and probably with quart-pots in their hands, but with an expression of unmistakable contentment and good-will. " Foh !" says my idealistic friend, "what vulgar details ! What good is there in

11

CA No. 6 SOLID
es 349 words in page
·2 words to line
6¾ = 25·26 square inches
vords to square inch
sures 41 × 23 = 943 ems
words to 1000 ems

S. W. GREEN'S SON
COMPOSITION, ELECTROTYPING, PRESSWORK
AND BINDING
74 and 76 Beekman Street
NEW YORK

So I am content to tell my simple story, without trying to make things seem better than they were ; dreading nothing, indeed, but falsity, which, in spite of one's best efforts, there is reason to dread. Falsehood is so easy, truth so difficult. The pencil is conscious of a delightful facility in drawing a griffin—the longer the claws, and the larger the wings, the better ; but that marvellous facility, which we mistook for genius, is apt to forsake us when we want to draw a real unexaggerated lion. Examine your words well, and you will find that even when you have no motive to be false, it is a very hard thing to say the exact truth, even about your own immediate feelings—much harder than to say something fine about them which is *not* the exact truth.

It is for this rare, precious quality of truthfulness that I delight in many Dutch paintings, which loftyminded people despise. I find a source of delicious sympathy in these faithful pictures of a monotonous homely existence, which has been the fate of so many more among my fellow-mortals than a life of pomp or of absolute indigence, of tragic suffering or of worldstirring actions. I turn without shrinking, from cloudborne angels, from prophets, sibyls, and heroic warriors, to an old woman bending over her flower-pot, or eating her solitary dinner, while the noonday light, softened, perhaps, by a screen of leaves, falls on her mob-cap, and just touches the rim of her spinningwheel, and her stone jug, and all those cheap, common things which are the precious necessaries of life to her ; or I turn to that village wedding, kept between four brown walls, where an awkward bridegroom opens

12

PICA O. S. No. 1 LEADED
32 lines 295 words in page
9·2 words to line
Page 3⅛ × 6¾ = 25·26 square inches
11·7 words to square inch
Page measures 41 × 23 = 943 ems
313 words to 1000 ems

S. W. GREEN'S SON
COMPOSITION, ELECTROTYPING, PRESSWOR
AND BINDING
74 and 76 Beekman Street
NEW YORK

So I am content to tell my simple story, without trying to make things seem better than they were; dreading nothing, indeed, but falsity, which, in spite of one's best efforts, there is reason to dread. Falsehood is so easy, truth so difficult. The pencil is conscious of a delightful facility in drawing a griffin—the longer the claws, and the larger the wings, the better; but that marvellous facility, which we mistook for genius, is apt to forsake us when we want to draw a real unexaggerated lion. Examine your words well, and you will find that even when you have no motive to be false, it is a very hard thing to say the exact truth, even about your own immediate feelings—much harder than to say something fine about them which is *not* the exact truth.

It is for this rare, precious quality of truthfulness that I delight in many Dutch paintings, which lofty-minded people despise. I find a source of delicious sympathy in these faithful pictures of a monotonous homely existence, which has been the fate of so many more among my fellow-mortals than a life of pomp or of absolute indigence, of tragic suffering or of world-stirring actions. I turn without shrinking, from cloud-borne angels, from prophets, sibyls, and heroic warriors, to an old woman bending over her flower-pot, or eating her solitary dinner, while the noonday light, softened, perhaps, by a screen of leaves, falls on her mob-cap, and just touches the rim of her spinning-wheel, and her stone jug, and all those cheap, common things which are the precious necessaries of life to her; or I turn to that village wedding, kept between four brown walls, where an awkward bridegroom opens the dance with a high-shouldered, broad-faced bride, while elderly and middle-aged friends look on, with very irregular noses and lips, and probably with quart-pots in their hands, but with an expression of unmistakable contentment and good-will. "Foh!" says my idealistic friend, "what vulgar details! What good

13

PICA O. S. No. 1 SOLID
38 lines 346 words in page
9·1 words to line
Page 3⅛ × 6⅞ = 25·26 square inches
13·7 words to square inch
Page measures 41 × 23 = 943 ems
367 words to 1000 ems

S. W. GREEN'S SON
COMPOSITION, ELECTROTYPING, PRESSWORK
AND BINDING
74 and 76 Beekman Street
NEW YORK

So I am content to tell my simple story, without trying to make things seem better than they were; dreading nothing, indeed, but falsity, which, in spite of one's best efforts, there is reason to dread. Falsehood is so easy, truth so difficult. The pencil is conscious of a delightful facility in drawing a griffin—the longer the claws, and the larger the wings, the better; but that marvellous facility, which we mistook for genius, is apt to forsake us when we want to draw a real unexaggerated lion. Examine your words well, and you will find that even when you have no motive to be false, it is a very hard thing to say the exact truth, even about your own immediate feelings—much harder than to say something fine about them which is *not* the exact truth.

It is for this rare, precious quality of truthfulness that I delight in many Dutch paintings, which lofty-minded people despise. I find a source of delicious sympathy in these faithful pictures of a monotonous homely existence, which has been the fate of so many more among my fellow-mortals than a life of pomp or of absolute indigence, of tragic suffering or of world-stirring actions. I turn without shrinking, from cloud-borne angels, from prophets, sibyls, and heroic warriors, to an old woman bending over her flower-pot, or eating her solitary dinner, while the noonday light, softened, perhaps, by a screen of leaves, falls on her mob-cap, and just touches the rim of her spinning-wheel, and her stone jug, and all those cheap, common things which are the precious necessaries of life to her; or I turn to that village wedding, kept between four brown walls, where an awkward bridegroom opens the dance with a high-shouldered, broad-faced bride, while elderly and middle-aged friends look on, with very irregular noses and lips, and probably with quart-pots in their hands, but with an expression of unmistakable contentment and good-will. "Foh!" says my idealistic friend, "what vulgar details! What good is there in taking all these pains to give an exact likeness

14

SMALL PICA No. 12 LEADED
37 lines 358 words in page
9.7 words to line
Page 3⅛ × 6⅞ = 25·26 square inches
14·2 words to square inch
Page measures 47 × 27 = 1269 ems
282 words to 1000 ems

S. W. GREEN'S S(
COMPOSITION, ELECTROTYPING,
AND BINDING
74 and 76 Beekman S
NEW YORK

Arranged from Fonts of Farmer, Little & Co.

PICA	SMALL PICA	LONG PRIMER	BOURGEOIS	BREVIER	MINION	NONPAREIL	AGATE	PEARL	PICA
1	1	1	1	1	1	2	2	2	1
2	2	2	3	3	3	4	5	5	2
3	3	4	4	5	5	6	7	7	3
4	5	5	6	6	7	8	9	10	4
5	6	6	7	8	9	10	12	12	5
6	7	7	8	9	10	12	14	15	6
7	8	9	10	11	12	14	16	17	7
8	9	10	11	12	14	16	19	20	8
9	10	11	13	14	15	18	21	22	9
10	12	12	14	15	17	20	23	25	10
11	13	14	15	17	19	22	26	27	11
12	14	15	17	18	21	24	28	30	12
13	15	16	18	20	22	26	30	32	13
14	16	17	20	21	24	28	32	35	14
15	17	19	21	23	26	30	35	37	15
16	19	20	22	24	28	32	37	40	16
17	20	21	24	26	29	34	39	42	17
18	21	22	25	27	31	36	42	45	18
19	22	24	27	29	33	38	44	47	19
20	23	25	28	30	34	40	46	50	20
21	24	26	29	32	36	42	49	52	21
22	26	27	31	33	38	44	51	55	22
23	27	29	32	35	40	46	53	57	23
24	28	30	34	36	41	48	56	60	24
25	29	31	35	38	43	50	58	62	25
26	30	32	36	40	45	52	60	64	26
27	31	33	38	41	46	54	63	67	27
28	32	35	39	43	48	56	65	69	28
29	34	36	41	44	50	58	67	72	29
30	35	37	42	46	52	60	70	74	30
31	36	38	43	47	53	62	72	77	31
32	37	40	45	49	55	64	74	79	32
33	38	41	46	50	57	66	77	82	33
34	39	42	48	52	58	68	79	84	34
35	41	43	49	53	60	70	81	87	35
36	42	45	50	55	62	72	84	89	36
37	43	46	52	56	64	74	86	92	37
38	44	47	53	58	65	76	88	94	38
39	45	48	55	59	67	78	90	97	39
40	46	50	56	61	69	80	93	99	40
41	48	51	57	62	71	82	95	102	41
42	49	52	59	64	72	84	97	104	42
43	50	53	60	65	74	86	100	107	43
44	51	55	62	67	76	88	102	109	44
45	52	56	63	68	77	90	104	112	45
46	53	57	64	70	79	92	107	114	46
47	55	58	66	71	81	94	109	117	47
48	56	60	67	73	83	96	111	119	48
49	57	61	69	74	84	98	114	122	49
50	58	62	70	76	86	100	116	124	50

LARGEST and BEST ORGANIZED COMPOSING
ROOMS in New York City.

ACTUAL AVERAGE PRODUCTION 3,500,000—4,000,000 EMS WEEKLY.

S. W. GREEN'S SON,

74 & 76 BEEKMAN STREET, NEW YORK CITY,

BOOK, NEWS and JOB

COMPOSITON and ELECTROTYPING,

PRESSWORK AND BINDING,

OF EVERY DESCRIPTION.

Every Facility for Correct and Rapid Work of all kinds in each department.

Prices as low as first-class workmanship will permit. SEND FOR ESTIMATES, which will be promptly furnished.

SCHOOL-BOOKS and MATHEMATICAL WORK
A Specialty.

So I am content to tell my simple story, without trying to make things seem better than they were; dreading nothing, indeed, but falsity, which, in spite of one's best efforts, there is reason to dread. Falsehood is so easy, truth so difficult. The pencil is conscious of a delightful facility in drawing a griffin—the longer the claws, and the larger the wings, the better; but that marvellous facility, which we mistook for genius, is apt to forsake us when we want to draw a real unexaggerated lion. Examine your words well, and you will find that even when you have no motive to be false, it is a very hard thing to say the exact truth, even about your own immediate feelings—much harder than to say something fine about them which is *not* the exact truth.

It is for this rare, precious quality of truthfulness that I delight in many Dutch paintings, which lofty-minded people despise. I find a source of delicious sympathy in these faithful pictures of a monotonous homely existence, which has been the fate of so many more among my fellow-mortals than a life of pomp or of absolute indigence, of tragic suffering or of world-stirring actions. I turn without shrinking, from cloud-borne angels, from prophets, sibyls, and heroic warriors, to an old woman bending over her flower-pot, or eating her solitary dinner, while the noonday light, softened, perhaps, by a screen of leaves, falls on her mob-cap, and just touches the rim of her spinning-wheel, and her stone jug, and all those cheap, common things which are the precious necessaries of life to her; or I turn to that village wedding, kept between four brown walls, where an awkward bridegroom opens the dance with a high-shouldered, broad-faced bride, while elderly and middle-aged friends look on, with very irregular noses and lips, and probably with quart-pots in their hands, but with an expression of unmistakable contentment and good-will. "Foh!" says my idealistic friend, "what vulgar details! What good is there in taking all these pains to give an exact likeness of old women and clowns? What a low phase of life! what clumsy, ugly people!"

But, bless us, things may be lovable that are not altogether handsome, I hope? I am not at all sure that the majority of the human race have not been ugly, and even among those "lords of their kind," the British, squat figures, ill-shapen nostrils, and dingy complexions are

15

SMALL PICA No. 12 SOLID
44 lines 423 words in page
9·6 words to line
Page 3⅛ × 6⅝ = 25·26 square inches
16·8 words to square inch
Page measures 47 × 27 = 1269 ems
111 words to 1000 ems

S. W. GREEN'S SON
COMPOSITION, ELECTROTYPING, PRESSWORK
AND BINDING
74 and 76 Beekman Street
NEW YORK

So I am content to tell my simple story, without trying to make things seem better than they were; dreading nothing, indeed, but falsity, which, in spite of one's best efforts, there is reason to dread. Falsehood is so easy, truth so difficult. The pencil is conscious of a delightful facility in drawing a griffin—the longer the claws, and the larger the wings, the better; but that marvellous facility, which we mistook for genius, is apt to forsake us when we want to draw a real un-exaggerated lion. Examine your words well, and you will find that even when you have no motive to be false, it is a very hard thing to say the exact truth, even about your own immediate feelings—much harder than to say something fine about them which is *not* the exact truth.

It is for this rare, precious quality of truthfulness that I delight in many Dutch paintings, which lofty-minded people despise. I find a source of delicious sympathy in these faith-ful pictures of a monotonous homely existence, which has been the fate of so many more among my fellow-mortals than a life of pomp or of absolute indigence, of tragic suffering or of world-stirring actions. I turn without shrinking, from cloud-borne angels, from prophets, sibyls, and heroic warriors, to an old woman bending over her flower-pot, or eating her solitary dinner, while the noonday light, softened, perhaps, by a screen of leaves, falls on her mob-cap, and just touches the rim of her spinning-wheel, and her stone jug, and all those cheap, common things which are the precious necessaries of life to her ; or I turn to that village wedding, kept between four brown walls, where an awkward bridegroom opens the dance with a high-shouldered, broad-faced bride, while elderly and middle-aged friends look on, with very irregular noses and lips, and prob-ably with quart-pots in their hands, but with an expression of unmistakable contentment and good-will. " Foh !" says my idealistic friend, " what vulgar details ! What good is there in taking all these pains to give an exact likeness of old women and clowns ? What a low phase of life ! what clumsy, ugly people !"

But, bless us, things may be lovable that are not altogether

16

SMALL PICA No. 9 LEADED
37 lines 384 words in page
10·4 words to line
Page 3⅛ × 6⅞ = 25·26 square inches
15·2 words to square inch
Page measures 47 × 27 = 1269 ems

S. W. GREEN'S SON
COMPOSITION, ELECTROTYPING, PRESSWORK
AND BINDING
74 and 76 Beekman Street

So I am content to tell my simple story, without trying to make things seem better than they were; dreading nothing, indeed, but falsity, which, in spite of one's best efforts, there is reason to dread. Falsehood is so easy, truth so difficult. The pencil is conscious of a delightful facility in drawing a griffin—the longer the claws, and the larger the wings, the better; but that marvellous facility, which we mistook for genius, is apt to forsake us when we want to draw a real unexaggerated lion. Examine your words well, and you will find that even when you have no motive to be false, it is a very hard thing to say the exact truth, even about your own immediate feelings—much harder than to say something fine about them which is *not* the exact truth.

It is for this rare, precious quality of truthfulness that I delight in many Dutch paintings, which lofty-minded people despise. I find a source of delicious sympathy in these faithful pictures of a monotonous homely existence, which has been the fate of so many more among my fellow-mortals than a life of pomp or of absolute indigence, of tragic suffering or of world-stirring actions. I turn without shrinking, from cloud-borne angels, from prophets, sibyls, and heroic warriors, to an old woman bending over her flower-pot, or eating her solitary dinner, while the noonday light, softened, perhaps, by a screen of leaves, falls on her mob-cap, and just touches the rim of her spinning-wheel, and her stone jug, and all those cheap, common things which are the precious necessaries of life to her; or I turn to that village wedding, kept between four brown walls, where an awkward bridegroom opens the dance with a high-shouldered, broad-faced bride, while elderly and middle-aged friends look on, with very irregular noses and lips, and probably with quart-pots in their hands, but with an expression of unmistakable contentment and good-will. "Foh!" says my idealistic friend, "what vulgar details! What good is there in taking all these pains to give an exact likeness of old women and clowns? What a low phase of life! what clumsy, ugly people!"

But, bless us, things may be lovable that are not altogether handsome, I hope? I am not at all sure that the majority of the human race have not been ugly, and even among those "lords of their kind," the British, squat figures, ill-shapen nostrils, and dingy complexions are not startling exceptions. Yet there is a great deal of family love among us. I have a friend or two whose class of features is such that the Apollo curl on the summit of their brows would be decidedly trying;

17

SMALL PICA No. 9 SOLID
44 lines 463 words in page
10·5 words to line
Page 3⅓ × 6⅝ = 25·26 square inches
18·3 words to square inch
Page measures 47 × 27 = 1269 ems
385 words to two ems

S. W. GREEN'S SON
COMPOSITION, ELECTROTYPING, PRESSWORK
AND BINDING
74 and 76 Beekman Street
NEW YORK

So I am content to tell my simple story, without trying
to make things seem better than they were; dreading
nothing, indeed, but falsity, which, in spite of one's best
efforts, there is reason to dread. Falsehood is so easy,
truth so difficult. The pencil is conscious of a delightful
facility in drawing a griffin—the longer the claws, and the
larger the wings, the better; but that marvellous facility,
which we mistook for genius, is apt to forsake us when
we want to draw a real unexaggerated lion. Examine
your words well, and you will find that even when you
have no motive to be false, it is a very hard thing to say
the exact truth, even about your own immediate feelings
—much harder than to say something fine about them
which is *not* the exact truth.

It is for this rare, precious quality of truthfulness that I
delight in many Dutch paintings, which lofty-minded people despise. I find a source of delicious sympathy in these
faithful pictures of a monotonous homely existence, which
has been the fate of so many more among my fellow-
mortals than a life of pomp or of absolute indigence, of
tragic suffering or of world-stirring actions. I turn without shrinking, from cloud-borne angels, from prophets,
sibyls, and heroic warriors, to an old woman bending over
her flower-pot, or eating her solitary dinner, while the
noonday light, softened, perhaps, by a screen of leaves,
falls on her mob-cap, and just touches the rim of her spinning-wheel, and her stone jug, and all those cheap, common things which are the precious necessaries of life to
her; or I turn to that village wedding, kept between four
brown walls, where an awkward bridegroom opens the
dance with a high-shouldered, broad-faced bride, while
elderly and middle-aged friends look on, with very irregular noses and lips, and probably with quart-pots in their
hands, but with an expression of unmistakable contentment
and good-will. "Foh!" says my idealistic friend, "what
vulgar details! What good is there in taking all these
pains to give an exact likeness of old women and clowns?

18

SMALL PICA O. S. No. 1 LEADED
37 lines 363 words in page
9·8 words to line
Page 3⅛ × 6⅝ = 25·26 square inches
14·4 words to square inch

S. W. GREEN'S SON
COMPOSITION, ELECTROTYPING, PRESSWORK
AND BINDING
74 and 76 Beekman Street

So I am content to tell my simple story, without trying to make things seem better than they were; dreading nothing, indeed, but falsity, which, in spite of one's best efforts, there is reason to dread. Falsehood is so easy, truth so difficult. The pencil is conscious of a delightful facility in drawing a griffin—the longer the claws, and the larger the wings, the better; but that marvellous facility, which we mistook for genius, is apt to forsake us when we want to draw a real unexaggerated lion. Examine your words well, and you will find that even when you have no motive to be false, it is a very hard thing to say the exact truth, even about your own immediate feelings —much harder than to say something fine about them which is *not* the exact truth.

It is for this rare, precious quality of truthfulness that I delight in many Dutch paintings, which lofty-minded people despise. I find a source of delicious sympathy in these faithful pictures of a monotonous homely existence, which has been the fate of so many more among my fellow-mortals than a life of pomp or of absolute indigence, of tragic suffering or of world-stirring actions. I turn without shrinking, from cloud-borne angels, from prophets, sibyls, and heroic warriors, to an old woman bending over her flower-pot, or eating her solitary dinner, while the noonday light, softened, perhaps, by a screen of leaves, falls on her mob-cap, and just touches the rim of her spinning-wheel, and her stone jug, and all those cheap, common things which are the precious necessaries of life to her; or I turn to that village wedding, kept between four brown walls, where an awkward bridegroom opens the dance with a high-shouldered, broad-faced bride, while elderly and middle-aged friends look on, with very irregular noses and lips, and probably with quart-pots in their hands, but with an expression of unmistakable contentment and good-will. "Foh!" says my idealistic friend, "what vulgar details! What good is there in taking all these pains to give an exact likeness of old women and clowns? What a low phase of life! what clumsy, ugly people!"

But, bless us, things may be lovable that are not altogether handsome, I hope? I am not at all sure that the majority of the human race have not been ugly, and even among those "lords of their kind," the British, squat figures, ill-shapen nostrils, and dingy complexions are not startling exceptions. Yet there is a great deal of family

SMALL PICA O. S. No. 1 SOLID
44 lines 434 words in page
9·9 words to line
Page 3½? × 6⅝ = 25·26 square inches
17·2 words to square inch
Page measures 47 × 27 = 1269 ems
342 words to 1000 ems

S. W. GREEN'S SON
COMPOSITION, ELECTROTYPING, PRESSWORK
AND BINDING
74 and 76 Beekman Street
NEW YORK

So I am content to tell my simple story, without trying to make things seem better than they were ; dreading nothing, indeed, but falsity, which, in spite of one's best efforts, there is reason to dread. Falsehood is so easy, truth so difficult. The pencil is conscious of a delightful facility in drawing a griffin—the longer the claws, and the larger the wings, the better; but that marvellous facility, which we mistook for genius, is apt to forsake us when we want to draw a real unexaggerated lion. Examine your words well, and you will find that even when you have no motive to be false, it is a very hard thing to say the exact truth, even about your own immediate feelings—much harder than to say something fine about them which is *not* the exact truth.

It is for this rare, precious quality of truthfulness that I delight in many Dutch paintings, which lofty-minded people despise. I find a source of delicious sympathy in these faithful pictures of a monotonous homely existence, which has been the fate of so many more among my fellow-mortals than a life of pomp or of absolute indigence, of tragic suffering or of world-stirring actions. I turn without shrinking, from cloud-borne angels, from prophets, sibyls, and heroic warriors, to an old woman bending over her flower-pot, or eating her solitary dinner, while the noonday light, softened, perhaps, by a screen of leaves, falls on her mob-cap, and just touches the rim of her spinning-wheel, and her stone jug, and all those cheap, common things which are the precious necessaries of life to her ; or I turn to that village wedding, kept between four brown walls, where an awkward bridegroom opens the dance with a high-shouldered, broad-faced bride, while elderly and middle-aged friends look on, with very irregular noses and lips, and probably with quart-pots in their hands, but with an expression of unmistakable contentment and good-will. "Foh!" says my idealistic friend, "what vulgar details! What good is there in taking all these pains to give an exact likeness of old women and clowns? What a low phase of life! what clumsy, ugly people!"

20

SMALL PICA O. S. No. 2 LEADED
37 lines 373 words in page
10 words to line
Page 3⅛ × 6¾ = 25·26 square inches
14·8 words to square inch
Page measures 47 × 27 = 1269 ems

S. W. GREEN'S SON
COMPOSITION, ELECTROTYPING, PRESSWORK
AND BINDING
74 and 76 Beekman Street

So I am content to tell my simple story, without trying
to make things seem better than they were ; dreading nothing, indeed, but falsity, which, in spite of one's best efforts,
there is reason to dread.   Falsehood is so easy, truth so
difficult.   The pencil is conscious of a delightful facility in
drawing a griffin—the longer the claws, and the larger the
wings, the better; but that marvellous facility, which we
mistook for genius, is apt to forsake us when we want to
draw a real unexaggerated lion.   Examine your words well,
and you will find that even when you have no motive to be
false, it is a very hard thing to say the exact truth, even
about your own immediate feelings—much harder than to
say something fine about them which is *not* the exact truth.
It is for this rare, precious quality of truthfulness that I
delight in many Dutch paintings, which lofty-minded people
despise.   I find a source of delicious sympathy in these
faithful pictures of a monotonous homely existence, which
has been the fate of so many more among my fellow-mortals
than a life of pomp or of absolute indigence, of tragic suffering or of world-stirring actions.   I turn without shrinking,
from cloud-borne angels, from prophets, sibyls, and heroic
warriors, to an old woman bending over her flower-pot, or
eating her solitary dinner, while the noonday light, softened,
perhaps, by a screen of leaves,,falls on her mob-cap, and just
touches the rim of her spinning-wheel, and her stone jug,
and all those cheap, common things which are the precious
necessaries of life to her ; or I turn to that village wedding,
kept between four brown walls, where an awkward bridegroom opens the dance with a high-shouldered, broad-faced
bride, while elderly and middle-aged friends look on, with
very irregular noses and lips, and probably with quart-pots
in their hands, but with an expression of⸱ unmistakable contentment and good-will.   " Foh !" says my idealistic friend,
"what vulgar details !   What good is there in taking all
these pains to give an exact likeness of old women and
clowns?   What a low phase of life! what clumsy, ugly
people !"
But, bless us, things may be lovable that are not altogether handsome, I hope?   I am not at all sure that the majority
of the human race have not been ugly, and even among those
"lords of their kind," the British, squat figures, ill-shapen
nostrils, and dingy complexions are not startling exceptions.
Yet there is a great deal of family love among us.   I have
a friend or two whose class of features is such that the

21

SMALL PICA O. S. No. 2 SOLID
44 lines 451 words in page
10·3 words to line
Page 3⅛ X 6⅝ = 25·26 square inches
17·9 words to square inch
Page measures 47 X 27 = 1269 ems

S. W. GREEN'S SON
COMPOSITION, ELECTROTYPING, PRESSWORK
AND BINDING
74 and 76 Beekman Street

So I am content to tell my simple story, without trying to make things seem better than they were; dreading nothing, indeed, but falsity, which, in spite of one's best efforts, there is reason to dread. Falsehood is so easy, truth so difficult. The pencil is conscious of a delightful facility in drawing a griffin— the longer the claws, and the larger the wings, the better; but that marvellous facility, which we mistook for genius, is apt to forsake us when we want to draw a real unexaggerated lion. Examine your words well, and you will find that even when you have no motive to be false, it is a very hard thing to say the exact truth, even about your own immediate feelings—much harder than to say something fine about them which is *not* the exact truth.

It is for this rare, precious quality of truthfulness that I delight in many Dutch paintings, which lofty-minded people despise. I find a source of delicious sympathy in these faithful pictures of a monotonous homely existence, which has been the fate of so many more among my fellow-mortals than a life of pomp or of absolute indigence, of tragic suffering or of world-stirring actions. I turn without shrinking, from cloud-borne angels, from prophets, sibyls, and heroic warriors, to an old woman bending over her flower-pot, or eating her solitary dinner, while the noonday light, softened, perhaps, by a screen of leaves, falls on her mob-cap, and just touches the rim of her spinning-wheel, and her stone jug, and all those cheap, common things which are the precious necessaries of life to her; or I turn to that village wedding, kept between four brown walls, where an awkward bridegroom opens the dance with a high-shouldered, broad-faced bride, while elderly and middle-aged friends look on, with very irregular noses and lips, and probably with quart-pots in their hands, but with an expression of unmistakable contentment and good-will. "Foh!" says my idealistic friend, "what vulgar details! What good is there in taking all these pains to give an exact likeness of old women and clowns? What a low phase of life! what clumsy, ugly people!"

But, bless us, things may be lovable that are not altogether handsome, I hope? I am not at all sure that the majority of the human race have not been ugly, and even among those "lords of their kind," the British, squat figures, ill-shapen

22

LONG PRIMER No. 13 LEADED
39 lines 418 words in page
10·7 words to line
Page 3|·] × 6¾ = 25·26 square inches
16·5 words to square inch
Page measures 50 × 29 = 1450 ems
288 words to 1000 ems

S. W. GREEN'S SON
COMPOSITION, ELECTROTYPING, PRESSWORK
AND BINDING
74 and 76 Beekman Street
NEW YORK

So I am content to tell my simple story, without trying to make things seem better than they were; dreading nothing, indeed, but falsity, which, in spite of one's best efforts, there is reason to dread. Falsehood is so easy, truth so difficult. The pencil is conscious of a delightful facility in drawing a griffin— the longer the claws, and the larger the wings, the better; but that marvellous facility, which we mistook for genius, is apt to forsake us when we want to draw a real unexaggerated lion. Examine your words well, and you will find that even when you have no motive to be false, it is a very hard thing to say the exact truth, even about your own immediate feelings—much harder than to say something fine about them which is *not* the exact truth.

It is for this rare, precious quality of truthfulness that I delight in many Dutch paintings, which lofty-minded people despise. I find a source of delicious sympathy in these faithful pictures of a monotonous homely existence, which has been the fate of so many more among my fellow-mortals than a life of pomp or of absolute indigence, of tragic suffering or of world-stirring actions. I turn without shrinking, from cloud-borne angels, from prophets, sibyls, and heroic warriors, to an old woman bending over her flower-pot, or eating her solitary dinner, while the noonday light, softened, perhaps, by a screen of leaves, falls on her mob-cap, and just touches the rim of her spinning-wheel, and her stone jug, and all those cheap, common things which are the precious necessaries of life to her; or I turn to that village wedding, kept between four brown walls, where an awkward bridegroom opens the dance with a high-shouldered, broad-faced bride, while elderly and middle-aged friends look on, with very irregular noses and lips, and probably with quart-pots in their hands, but with an expression of unmistakable contentment and good-will. "Foh!" says my idealistic friend, "what vulgar details! What good is there in taking all these pains to give an exact likeness of old women and clowns? What a low phase of life! what clumsy, ugly people!"

But, bless us, things may be lovable that are not altogether handsome, I hope? I am not at all sure that the majority of the human race have not been ugly, and even among those "lords of their kind," the British, squat figures, ill-shapen nostrils, and dingy complexions are not startling exceptions. Yet there is a great deal of family love among us. I have a friend or two whose class of features is such that the Apollo curl on the summit of their brows would be decidedly trying; yet, to my certain knowledge, tender hearts have beaten for them, and their miniatures—flattering, but still not lovely—are kissed in secret by motherly lips. I have seen many an excellent matron who could never in her best days have been handsome, and yet

23

LONG PRIMER No. 13 SOLID
47 lines 508 words in page
10·8 words to line
Page 3⅞ × 6⅝ = 25·26 square inches
20 words to square inch
Page measures 50 × 29 = 1450 ems
350 words to 1000 ems

S. W. GREEN'S SON
COMPOSITION, ELECTROTYPING, PRESSWORK
AND BINDING
74 and 76 Beekman Street
NEW YORK

So I am content to tell my simple story, without trying to make things seem better than they were; dreading nothing, indeed, but falsity, which, in spite of one's best efforts, there is reason to dread. Falsehood is so easy, truth so difficult. The pencil is conscious of a delightful facility in drawing a griffin—the longer the claws, and the larger the wings, the better; but that marvellous facility, which we mistook for genius, is apt to forsake us when we want to draw a real unexaggerated lion. Examine your words well, and you will find that even when you have no motive to be false, it is a very hard thing to say the exact truth, even about your own immediate feelings—much harder than to say something fine about them which is *not* the exact truth.

It is for this rare, precious quality of truthfulness that I delight in many Dutch paintings, which lofty-minded people despise. I find a source of delicious sympathy in these faithful pictures of a monotonous homely existence, which has been the fate of so many more among my fellow-mortals than a life of pomp or of absolute indigence, of tragic suffering or of world-stirring actions. I turn without shrinking, from cloud-borne angels, from prophets, sibyls, and heroic warriors, to an old woman bending over her flower-pot, or eating her solitary dinner, while the noonday light, softened, perhaps, by a screen of leaves, falls on her mob-cap, and just touches the rim of her spinning-wheel, and her stone jug, and all those cheap, common things which are the precious necessaries of life to her; or I turn to that village wedding, kept between four brown walls, where an awkward bridegroom opens the dance with a high-shouldered, broad-faced bride, while elderly and middle-aged friends look on, with very irregular noses and lips, and probably with quart-pots in their hands, but with an expression of unmistakable contentment and good-will. "Foh!" says my idealistic friend, "what vulgar details! What good is there in taking all these pains to give an exact likeness of old women and clowns? What a low phase of life! what clumsy, ugly people!"

But, bless us, things may be lovable that are not altogether handsome, I hope? I am not at all sure that the majority of the human race have not been ugly, and even among those "lords of their kind," the British, squat figures, ill-shapen nostrils, and dingy complexions are not startling exceptions. Yet there is a great deal of family love among us. I have a friend or two whose class of features

24

LONG PRIMER No. 9 LEADED
39 lines 447 words in page
11·5 words to line
Page 3⅛ × 6¾ = 25·26 square inches
17·7 words to square inch
Page measures 50 × 29 = 1450 ems
308 words to 1000 ems

S. W. GREEN'S SON
COMPOSITION, ELECTROTYPING, PRESSWORK
AND BINDING
74 and 76 Beekman Street
NEW YORK

So I am content to tell my simple story, without trying to make things seem better than they were; dreading nothing, indeed, but falsity, which, in spite of one's best efforts, there is reason to dread. Falsehood is so easy, truth so difficult. The pencil is conscious of a delightful facility in drawing a griffin—the longer the claws, and the larger the wings, the better; but that marvellous facility, which we mistook for genius, is apt to forsake us when we want to draw a real unexaggerated lion. Examine your words well, and you will find that even when you have no motive to be false, it is a very hard thing to say the exact truth, even about your own immediate feelings—much harder than to say something fine about them which is *not* the exact truth.

It is for this rare, precious quality of truthfulness that I delight in many Dutch paintings, which lofty-minded people despise. I find a source of delicious sympathy in these faithful pictures of a monotonous homely existence, which has been the fate of so many more among my fellow-mortals than a life of pomp or of absolute indigence, of tragic suffering or of world-stirring actions. I turn without shrinking, from cloud-borne angels, from prophets, sibyls, and heroic warriors, to an old woman bending over her flower-pot, or eating her solitary dinner, while the noonday light, softened, perhaps, by a screen of leaves, falls on her mob-cap, and just touches the rim of her spinning-wheel, and her stone jug, and all those cheap, common things which are the precious necessaries of life to her; or I turn to that village wedding, kept between four brown walls, where an awkward bridegroom opens the dance with a high-shouldered, broad-faced bride, while elderly and middle-aged friends look on, with very irregular noses and lips, and probably with quart-pots in their hands, but with an expression of unmistakable contentment and good-will. "Foh!" says my idealistic friend, "what vulgar details! What good is there in taking all these pains to give an exact likeness of old women and clowns? What a low phase of life! what clumsy, ugly people!"

But, bless us, things may be lovable that are not altogether handsome, I hope? I am not at all sure that the majority of the human race have not been ugly, and even among those "lords of their kind," the British, squat figures, ill-shapen nostrils, and dingy complexions are not startling exceptions. Yet there is a great deal of family love among us. I have a friend or two whose class of features is such that the Apollo curl on the summit of their brows would be decidedly trying; yet, to my certain knowledge, tender hearts have beaten for them, and their miniatures—flattering, but still not lovely— are kissed in secret by motherly lips. I have seen many an excellent matron who could never in her best days have been handsome, and yet she had a packet of yellow love-letters in a private drawer, and sweet children showered kisses on her sallow cheeks. And I believe there have been plenty of young heroes, of middle stature and feeble

25

LONG PRIMER No. 9 SOLID
47 lines 544 words in page
11·6 words to line
Page 3½½ × 6⅝ = 25·26 square inches
21·6 words to square inch
Page measures 50 × 29 = 1450 ems
375 words to 1000 ems

S. W. GREEN'S SON
COMPOSITION, ELECTROTYPING, PRESSWORK
AND BINDING
74 and 76 Beekman Street
NEW YORK

So I am content to tell my simple story, without trying to make things seem better than they were; dreading nothing, indeed, but falsity, which, in spite of one's best efforts, there is reason to dread. Falsehood is so easy, truth so difficult. The pencil is conscious of a delightful facility in drawing a griffin—the longer the claws, and the larger the wings, the better; but that marvellous facility, which we mistook for genius, is apt to forsake us when we want to draw a real unexaggerated lion. Examine your words well, and you will find that even when you have no motive to be false, it is a very hard thing to say the exact truth, even about your own immediate feelings—much harder than to say something fine about them which is *not* the exact truth.

It is for this rare, precious quality of truthfulness that I delight in many Dutch paintings, which lofty-minded people despise. I find a source of delicious sympathy in these faithful pictures of a monotonous homely existence, which has been the fate of so many more among my fellow-mortals than a life of pomp or of absolute indigence, of tragic suffering or of world-stirring actions. I turn without shrinking, from cloud-borne angels, from prophets, sibyls, and heroic warriors, to an old woman bending over her flower-pot, or eating her solitary dinner, while the noonday light, softened, perhaps, by a screen of leaves, falls on her mob-cap, and just touches the rim of her spinning-wheel, and her stone jug, and all those cheap, common things which are the precious necessaries of life to her; or I turn to that village wedding, kept between four brown walls, where an awkward bridegroom opens the dance with a high-shouldered, broad-faced bride, while elderly and middle-aged friends look on, with very irregular noses and lips, and probably with quart-pots in their hands, but with an expression of unmistakable contentment and good-will. "Foh!" says my idealistic friend, "what vulgar details! What good is there in taking all these pains to give an exact likeness of old women and clowns? What a low phase of life! what clumsy, ugly people!"

But, bless us, things may be lovable that are not altogether handsome, I hope? I am not at all sure that the majority of the human race have not been ugly, and even among those "lords of their kind," the British, squat figures, ill-shapen nostrils, and dingy complexions are not startling exceptions. Yet there is a great deal of family love among us. I have a friend or two whose

26

LONG PRIMER O. S. No. 2 LEADED
39 lines 444 words in page
11·4 words to line
Page 3$\frac{13}{16}$ × 6$\frac{3}{16}$ = 25·16 square inches
17·6 words to square inch
Page measures 50 × 29 = 1450 ems
306 words to 1000 ems

S. W. GREEN'S SON
COMPOSITION, ELECTROTYPING, PRESSWORK
AND BINDING
74 and 76 Beekman Street
NEW YORK

So I am content to tell my simple story, without trying to make things seem better than they were; dreading nothing, indeed, but falsity, which, in spite of one's best efforts, there is reason to dread. Falsehood is so easy, truth so difficult. The pencil is conscious of a delightful facility in drawing a griffin—the longer the claws, and the larger the wings, the better; but that marvellous facility, which we mistook for genius, is apt to forsake us when we want to draw a real unexaggerated lion. Examine your words well, and you will find that even when you have no motive to be false, it is a very hard thing to say the exact truth, even about your own immediate feelings—much harder than to say something fine about them which is *not* the exact truth.

It is for this rare, precious quality of truthfulness that I delight in many Dutch paintings, which lofty-minded people despise. I find a source of delicious sympathy in these faithful pictures of a monotonous homely existence, which has been the fate of so many more among my fellow-mortals than a life of pomp or of absolute indigence, of tragic suffering or of world-stirring actions. I turn without shrinking, from cloud-borne angels, from prophets, sibyls, and heroic warriors, to an old woman bending over her flower-pot, or eating her solitary dinner, while the noonday light, softened, perhaps, by a screen of leaves, falls on her mob-cap, and just touches the rim of her spinning-wheel, and her stone jug, and all those cheap, common things which are the precious necessaries of life to her; or I turn to that village wedding, kept between four brown walls, where an awkward bridegroom opens the dance with a high-shouldered, broad-faced bride, while elderly and middle-aged friends look on, with very irregular noses and lips, and probably with quart-pots in their hands, but with an expression of unmistakable contentment and good-will. "Foh!" says my idealistic friend, "what vulgar details! What good is there in taking all these pains to give an exact likeness of old women and clowns? What a low phase of life! what clumsy, ugly people!"

But, bless us, things may be lovable that are not altogether handsome, I hope? I am not at all sure that the majority of the human race have not been ugly, and even among those "lords of their kind," the British, squat figures, ill-shapen nostrils, and dingy complexions are not startling exceptions. Yet there is a great deal of family love among us. I have a friend or two whose class of features is such that the Apollo curl on the summit of their brows would be decidedly trying; yet, to my certain knowledge, tender hearts have beaten for them, and their miniatures—flattering, but still not lovely—are kissed in secret by motherly lips. I have seen many an excellent matron who could never in her best days have been handsome, and yet she had a packet of yellow love-letters in a private drawer, and sweet children showered kisses on her sallow cheeks. And I believe there have been

27

LONG PRIMER O. S. No. 2 SOLID
47 lines 535 words in page
11·4 words to line
Page 3⅛ × 6⅝ = 25·26 square inches
21·2 words to square inch
Page measures 50 × 29 = 1450 ems
360 words to 1000 ems

S. W. GREEN'S SON
COMPOSITION, ELECTROTYPING, PRESSWORK
AND BINDING
74 and 76 Beekman Street
NEW YORK

So I am content to tell my simple story, without trying to make things seem better than they were ; dreading nothing, indeed, but falsity, which, in spite of one's best efforts, there is reason to dread. Falsehood is so easy, truth so difficult. The pencil is conscious of a delightful facility in drawing a griffin—the longer the claws, and the larger the wings, the better ; but that marvellous facility, which we mistook for genius, is apt to forsake us when we want to draw a real un-exaggerated lion. Examine your words well, and you will find that even when you have no motive to be false, it is a very hard thing to say the exact truth, even about your own imme-diate feelings—much harder than to say something fine about them which is *not* the exact truth.

It is for this rare, precious quality of truthfulness that I delight in many Dutch paintings, which lofty-minded people despise. I find a source of delicious sympathy in these faith-ful pictures of a monotonous homely existence, which has been the fate of so many more among my fellow-mortals than a life of pomp or of absolute indigence, of tragic suffering or of world-stirring actions. I turn without shrinking, from cloud-borne angels, from prophets, sibyls, and heroic warriors, to an old woman bending over her flower-pot, or eating her solitary dinner, while the noonday light, softened, perhaps, by a screen of leaves, falls on her mob-cap, and just touches the rim of her spinning-wheel, and her stone jug, and all those cheap, com-mon things which are the precious necessaries of life to her ; or I turn to that village wedding, kept between four brown walls, where an awkward bridegroom opens the dance with a high-shouldered, broad-faced bride, while elderly and middle-aged friends look on, with very irregular noses and lips, and probably with quart-pots in their hands, but with an expression of unmistakable contentment and good-will. "Foh !" says my idealistic friend, "what vulgar details ! What good is there in taking all these pains to give an exact likeness of old women and clowns ? What a low phase of life ! what clumsy, ugly people !"

But, bless us, things may be lovable that are not altogether handsome, I hope ? I am not at all sure that the majority of the human race have not been ugly, and even among those

28

LONG PRIMER O. S. No. 3 LEADED
39 lines 408 words in page
10·5 words to line
Page 3½ × 6⅜ = 25·26 square inches
16·2 words to square inch
Page measures 50 × 29 = 1450 ems
281 words to 1000 ems

S. W. GREEN'S SON
COMPOSITION, ELECTROTYPING, PRESSWORK
AND BINDING
74 and 76 Beekman Street
NEW YORK

So I am content to tell my simple story, without trying to make things seem better than they were ; dreading nothing, indeed, but falsity, which, in spite of one's best efforts, there is reason to dread. Falsehood is so easy, truth so difficult. The pencil is conscious of a delightful facility in drawing a griffin—the longer the claws, and the larger the wings, the better ; but that marvellous facility which we mistook for genius, is apt to forsake us when we want to draw a real unexaggerated lion. Examine your words well, and you will find that even when you have no motive to be false, it is a very hard thing to say the exact truth, even about your own immediate feelings—much harder than to say something fine about them which is *not* the exact truth.

It is for this rare, precious quality of truthfulness that I delight in many Dutch paintings, which lofty-minded people despise. I find a source of delicious sympathy in these faithful pictures of a monotonous homely existence, which has been the fate of so many more among my fellow-mortals than a life of pomp or of absolute indigence, of tragic suffering or of world-stirring actions. I turn without shrinking, from cloud-borne angels, from prophets, sibyls, and heroic warriors, to an old woman bending over her flower-pot, or eating her solitary dinner, while the noonday light, softened, perhaps, by a screen of leaves, falls on her mob-cap, and just touches the rim of her spinning-wheel, and her stone jug, and all those cheap, common things which are the precious necessaries of life to her ; or I turn to that village wedding, kept between four brown walls, where an awkward bridegroom opens the dance with a high-shouldered, broad-faced bride, while elderly and middle-aged friends look on, with very irregular noses and lips, and probably with quart-pots in their hands, but with an expression of unmistakable contentment and good-will. "Foh !" says my idealistic friend, "what vulgar details ! What good is there in taking all these pains to give an exact likeness of old women and clowns ? What a low phase of life ! what clumsy, ugly people !"

But, bless us, things may be lovable that are not altogether handsome, I hope ? I am not at all sure that the majority of the human race have not been ugly, and even among those "lords of their kind," the British, squat figures, ill-shapen nostrils, and dingy complexions are not startling exceptions. Yet there is a great deal of family love among us. I have a friend or two whose class of features is such that the Apollo curl on the summit of their brows would be decidedly trying ; yet, to my certain knowledge, tender hearts have beaten for them, and their miniatures—flattering, but still not lovely— are kissed in secret by motherly lips. I have seen many an

29

LONG PRIMER O. S. No. 3 SOLID
47 lines 494 words in page
10·5 words to line
Page 3⅛ × 6¾ = 25·26 square inches
19·6 words to square inch
Page measures 50 × 29 = 1450 ems
340 words to 1000 ems

S. W. GREEN'S SON
COMPOSITION, ELECTROTYPING, PRESSWORK
AND BINDING
74 and 76 Beekman Street
NEW YORK

So I am content to tell my simple story, without trying to make things seem better than they were; dreading nothing, indeed, but falsity, which, in spite of one's best efforts, there is reason to dread. Falsehood is so easy, truth so difficult. The pencil is conscious of a delightful facility in drawing a griffin—the longer the claws, and the larger the wings, the better; but that marvellous facility, which we mistook for genius, is apt to forsake us when we want to draw a real unexaggerated lion. Examine your words well, and you will find that even when you have no motive to be false, it is a very hard thing to say the exact truth, even about your own immediate feelings—much harder than to say something fine about them which is *not* the exact truth.

It is for this rare, precious quality of truthfulness that I delight in many Dutch paintings, which lofty-minded people despise. I find a source of delicious sympathy in these faithful pictures of a monotonous homely existence, which has been the fate of so many more among my fellow-mortals than a life of pomp or of absolute indigence, of tragic suffering or of world-stirring actions. I turn without shrinking, from cloud-borne angels, from prophets, sibyls, and heroic warriors, to an old woman bending over her flower-pot, or eating her solitary dinner, while the noonday light, softened, perhaps, by a screen of leaves, falls on her mob-cap, and just touches the rim of her spinning-wheel, and her stone jug, and all those cheap, common things which are the precious necessaries of life to her; or I turn to that village wedding, kept between four brown walls, where an awkward bridegroom opens the dance with a high-shouldered, broad-faced bride, while elderly and middle-aged friends look on, with very irregular noses and lips, and probably with quart-pots in their hands, but with an expression of unmistakable contentment and good-will. "Foh!" says my idealistic friend, "what vulgar details! What good is there in taking all these pains to give an exact likeness of old women and clowns? What a low phase of life! what clumsy, ugly people!"

But, bless us, things may be lovable that are not altogether handsome, I hope? I am not at all sure that the majority of the human race have not been ugly, and even among those "lords of their kind," the British, squat figures, ill-shapen nostrils, and dingy complexions are not startling exceptions. Yet there is a great deal of family love among us. I have a friend or two whose class of features is such that the Apollo curl on the summit of their brows would be decidedly trying; yet, to my certain knowledge, tender hearts have beaten for them, and their miniatures—flattering, but still not lovely—are kissed in secret by motherly lips. I have seen many an excellent matron who could never in her best days have been handsome, and yet she had a packet of yellow love-letters in a private drawer, and sweet children showered kisses on her sallow cheeks. And I believe there have been plenty of young

· 30

BOURGEOIS No. 12 LEADED
44 lines 538 words in page
12½ words to line
Page 3½⅞ × 6¾ = 25·26 square inches
21·3 words to square inch
Page measures 57 × 32 = 1824 ems
295 words to 1000 ems

S. W. GREEN'S SON
COMPOSITION, ELECTROTYPING, PRESSWORK
AND BINDING
74 and 76 Beekman Street
NEW YORK

So I am content to tell my simple story, without trying to make things seem better than they were; dreading nothing, indeed, but falsity, which, in spite of one's best efforts, there is reason to dread. Falsehood is so easy, truth so difficult. The pencil is conscious of a delightful facility in drawing a griffin—the longer the claws, and the larger the wings, the better; but that marvellous facility, which we mistook for genius, is apt to forsake us when we want to draw a real unexaggerated lion. Examine your words well, and you will find that even when you have no motive to be false, it is a very hard thing to say the exact truth, even about your own immediate feelings—much harder than to say something fine about them which is *not* the exact truth.

It is for this rare, precious quality of truthfulness that I delight in many Dutch paintings, which lofty-minded people despise. I find a source of delicious sympathy in these faithful pictures of a monotonous homely existence, which has been the fate of so many more among my fellow-mortals than a life of pomp or of absolute indigence, of tragic suffering or of world-stirring actions. I turn without shrinking, from cloud-borne angels, from prophets, sibyls, and heroic warriors, to an old woman bending over her flower-pot, or eating her solitary dinner, while the noonday light, softened, perhaps, by a screen of leaves, falls on her mob-cap, and just touches the rim of her spinning-wheel, and her stone jug, and all those cheap, common things which are the precious necessaries of life to her; or I turn to that village wedding, kept between four brown walls, where an awkward bridegroom opens the dance with a high-shouldered, broad-faced bride, while elderly and middle-aged friends look on, with very irregular noses and lips, and probably with quart-pots in their hands, but with an expression of unmistakable contentment and good-will. "Foh!" says my idealistic friend, "what vulgar details! What good is there in taking all these pains to give an exact likeness of old women and clowns? What a low phase of life! what clumsy, ugly people!"

But, bless us, things may be lovable that are not altogether handsome, I hope? I am not at all sure that the majority of the human race have not been ugly, and even among those "lords of their kind," the British, squat figures, ill-shapen nostrils, and dingy complexions are not startling exceptions. Yet there is a great deal of family love among us. I have a friend or two whose class of features is such that the Apollo curl on the summit of their brows would be decidedly trying; yet, to my certain knowledge, tender hearts have beaten for them, and their miniatures—flattering, but still not lovely—are kissed in secret by motherly lips. I have seen many an excellent matron who could never in her best days have been handsome, and yet she had a packet of yellow love-letters in a private drawer, and sweet children showered kisses on her sallow cheeks. And I believe there have been plenty of young heroes, of middle stature and feeble beards, who have felt quite sure they could never love anything more insignificant than a Diana, and yet have found themselves in middle life happily settled with a wife who waddles. Yes! thank God; human feeling is like the mighty rivers that bless the earth; it does not wait for beauty—it flows with resistless force, and brings beauty with it.

All honor and reverence to the divine beauty of form! Let us cultivate it to the utmost in men, women, and children—in our gardens and in our houses; but let us love that other beauty, too, which lies in no secret of proportion, but in the secret of deep human sympathy. Paint

31

BOURGEOIS No. 12 SOLID
54 lines 658 words in page
12·2 words to line
Page 3½ × 6⅝ = 25·26 square inches
26 words to square inch
Page measures 57 × 32 = 1824 ems

S. W. GREEN'S SON
COMPOSITION, ELECTROTYPING PRESSWORK
AND BINDING
74 and 76 Beekman Street
NEW YORK

So I am content to tell my simple story, without trying to make things seem better than they were; dreading nothing, indeed, but falsity, which, in spite of one's best efforts, there is reason to dread. Falsehood is so easy, truth so difficult. The pencil is conscious of a delightful facility in drawing a griffin—the longer the claws, and the larger the wings, the better; but that marvellous facility, which we mistook for genius, is apt to forsake us when we want to draw a real unexaggerated lion. Examine your words well, and you will find that even when you have no motive to be false, it is a very hard thing to say the exact truth, even about your own immediate feelings—much harder than to say something fine about them which is *not* the exact truth.

It is for this rare, precious quality of truthfulness that I delight in many Dutch paintings, which lofty-minded people despise. I find a source of delicious sympathy in these faithful pictures of a monotonous homely existence, which has been the fate of so many more among my fellow-mortals than a life of pomp or of absolute indigence, of tragic suffering or of world-stirring actions. I turn without shrinking, from cloud-borne angels, from prophets, sibyls, and heroic warriors, to an old woman bending over her flower-pot, or eating her solitary dinner, while the noonday light, softened, perhaps, by a screen of leaves, falls on her mob-cap, and just touches the rim of her spinning-wheel, and her stone jug, and all those cheap, common things which are the precious necessaries of life to her; or I turn to that village wedding, kept between four brown walls, where an awkward bridegroom opens the dance with a high-shouldered, broad-faced bride, while elderly and middle-aged friends look on, with very irregular noses and lips, and probably with quart-pots in their hands, but with an expression of unmistakable contentment and good-will. " Foh !" says my idealistic friend, "what vulgar details! What good is there in taking all these pains to give an exact likeness of old women and clowns ? What a low phase of life ! what clumsy, ugly people !"

But, bless us, things may be lovable that are not altogether handsome, I hope? I am not at all sure that the majority of the human race have not been ugly, and even among those "lords of their kind," the British, squat figures, ill-shapen nostrils, and dingy complexions are not startling exceptions. Yet there is a great deal of family love among us. I have a friend or two whose class of features is such that the Apollo curl on the summit of their brows would be decidedly trying; yet, to my certain knowledge, tender hearts have beaten for them, and their miniatures—flattering, but still not lovely—are kissed in secret by motherly lips. I have seen many an excellent matron who could never in her best days have been handsome, and yet she had a packet of yellow love-letters in a private drawer, and sweet chil-

32

BOURGEOIS O. S. No. 3 LEADED
44 lines 523 words in page
11·9 words to line
Page 3⅛ × 6⅝ = 25·26 square inches
20·7 words to square inch
Page measures 57 × 32 = 1824 ems
287 words to 1000 ems

S. W. GREEN'S SON
COMPOSITION, ELECTROTYPING, PRESSWORK
AND BINDING
74 and 76 Beekman Street

So I am content to tell my simple story, without trying to make things seem better than they were; dreading nothing, indeed, but falsity, which, in spite of one's best efforts, there is reason to dread. Falsehood is so easy, truth so difficult. The pencil is conscious of a delightful facility in drawing a griffin—the longer the claws, and the larger the wings, the better; but that marvellous facility, which we mistook for genius, is apt to forsake us when we want to draw a real unexaggerated lion. Examine your words well, and you will find that even when you have no motive to be false, it is a very hard thing to say the exact truth, even about your own immediate feelings—much harder than to say something fine about them which is *not* the exact truth.

It is for this rare, precious quality of truthfulness that I delight in many Dutch paintings, which lofty-minded people despise. I find a source of delicious sympathy in these faithful pictures of a monotonous homely existence, which has been the fate of so many more among my fellow-mortals than a life of pomp or of absolute indigence, of tragic suffering or of world-stirring actions. I turn without shrinking, from cloud-borne angels, from prophets, sibyls, and heroic warriors, to an old woman bending over her flower-pot, or eating her solitary dinner, while the noonday light, softened, perhaps, by a screen of leaves, falls on her mob-cap, and just touches the rim of her spinning-wheel, and her stone jug, and all those cheap, common things which are the precious necessaries of life to her; or I turn to that village wedding, kept between four brown walls, where an awkward bridegroom opens the dance with a high-shouldered, broad-faced bride, while elderly and middle-aged friends look on, with very irregular noses and lips, and probably with quart-pots in their hands, but with an expression of unmistakable contentment and good-will. " Foh!" says my idealistic friend, "what vulgar details! What good is there in taking all these pains to give an exact likeness of old women and clowns? What a low phase of life! what clumsy, ugly people!"

But, bless us, things may be lovable that are not altogether handsome, I hope? I am not at all sure that the majority of the human race have not been ugly, and even among those "lords of their kind," the British, squat figures, ill-shapen nostrils, and dingy complexions are not startling exceptions. Yet there is a great deal of family love among us. I have a friend or two whose class of features is such that the Apollo curl on the summit of their brows would be decidedly trying; yet, to my certain knowledge, tender hearts have beaten for them, and their miniatures—flattering, but still not lovely—are kissed in secret by motherly lips. I have seen many an excellent matron who could never in her best days have been handsome, and yet she had a packet of yellow love-letters in a private drawer, and sweet children showered kisses on her sallow cheeks. And I believe there have been plenty of young heroes, of middle stature and feeble beards, who have felt quite sure they could never love anything more insignificant than a Diana, and yet have found themselves in middle life happily settled with a wife who waddles. Yes! thank God; human feeling is like the mighty rivers that bless the earth; it does not wait for beauty—it flows with resistless force, and brings beauty with it.

All honor and reverence to the divine beauty of form! Let us cultivate it to the utmost in men, women, and children—in our gardens and in our houses; but let us love that other beauty, too, which lies.

33

BOURGEOIS O. S. No. 3 SOLID
54 lines 644 words in page
12 words in line
Page 3¾ X 6⅜ = 25·26 square inches
25·5 words to square inch
Page measures 57 X 32 = 1824 ems

S. W. GREEN'S SCN
COMPOSITION, ELECTROTYPING, PRESSWORK
AND BINDING
74 and 76 Beekman Street

So I am content to tell my simple story, without trying to make things seem better than they were; dreading nothing, indeed, but falsity, which, in spite of one's best efforts, there is reason to dread. Falsehood is so easy, truth so difficult. The pencil is conscious of a delightful facility in drawing a griffin—the longer the claws, and the larger the wings, the better; but that marvellous facility, which we mistook for genius, is apt to forsake us when we want to draw a real unexaggerated lion. Examine your words well, and you will find that even when you have no motive to be false, it is a very hard thing to say the exact truth, even about your own immediate feelings—much harder than to say something fine about them which is *not* the exact truth.

It is for this rare, precious quality of truthfulness that I delight in many Dutch paintings, which lofty-minded people despise. I find a source of delicious sympathy in these faithful pictures of a monotonous homely exist-ence, which has been the fate of so many more among my fellow-mortals than a life of pomp or of absolute indigence, of tragic suffering or of world-stirring actions. I turn without shrinking, from cloud-borne angels, from prophets, sibyls, and heroic warriors, to an old woman bending over her flower-pot, or eating her solitary dinner, while the noonday light, softened, perhaps, by a screen of leaves, falls on her mob-cap, and just touches the rim of her spinning-wheel, and her stone jug, and all those cheap, common things which are the precious necessaries of life to her; or I turn to that village wedding, kept between four brown walls, where an awkward bride-groom opens the dance with a high-shouldered, broad-faced bride, while elderly and middle-aged friends look on, with very irregular noses and lips, and probably with quart-pots in their hands, but with an expression of unmistakable contentment and good-will. "Foh!" says my idealistic friend, "what vulgar details! What good is there in taking all those pains to give an exact likeness of old women and clowns? What a low phase of life! what clumsy, ugly people!"

But, bless us, things may be lovable that are not altogether handsome, I hope? I am not at all sure that the majority of the human race have not been ugly, and even among those "lords of their kind," the British, squat figures, ill-shapen nostrils, and dingy complexions are not startling excep-tions. Yet there is a great deal of family love among us. I find a friend or two whose class of features is such that the Apollo curl on the summit of their brows would be decidedly trying; yet, to my certain knowledge, tender hearts have beaten for them, and their miniatures—flattering, but still not lovely—are kissed in secret by motherly lips. I have seen many an excellent matron who could never in her best days have been hand-some, and yet she had a packet of yellow love-letters in a private drawer, and sweet children showered kisses on her sallow cheeks. And I believe there have been plenty of young heroes, of middle stature and feeble beards, who have felt quite sure they could never love anything more insignificant than a Diana, and yet have found themselves in middle life happily settled with a wife who waddles. Yes! thank God; human feeling is like the

34

BREVIER No. 12 LEADED
46 lines 583 words in page
12·7 words to line
Page 3½ × 6⅝ = 25·26 square inches
23·1 words to square inch
Page measures 61 × 35 = 2135 ems
271 words to 1000 ems

S. W. GREEN'S SON
COMPOSITION, ELECTROTYPING, PRESSWORK
AND BINDING
74 and 76 Beekman Street
NEW YORK

So I am content to tell my simple story, without trying to make things seem better than they were; dreading nothing, indeed, but falsity, which, in spite of one's best efforts, there is reason to dread. Falsehood is so easy, truth so difficult. The pencil is conscious of a delightful facility in drawing a griffin—the longer the claws, and the larger the wings, the better; but that marvellous facility which we mistook for genius, is apt to forsake us when we want to draw a real unexaggerated lion. Examine your words well, and you will find that even when you have no motive to be false, it is a very hard thing to say the exact truth, even about your own immediate feelings—much harder than to say something fine about them which is *not* the exact truth.

It is for this rare, precious quality of truthfulness that I delight in many Dutch paintings, which lofty-minded people despise. I find a source of delicious sympathy in these faithful pictures of a monotonous homely existence, which has been the fate of so many more among my fellow-mortals than a life of pomp or of absolute indigence, of tragic suffering or of world-stirring actions. I turn without shrinking, from cloud-borne angels, from prophets, sibyls, and heroic warriors, to an old woman bending over her flower-pot, or eating her solitary dinner, while the noonday light, softened, perhaps, by a screen of leaves, falls on her mob-cap, and just touches the rim of her spinning-wheel, and her stone jug, and all those cheap, common things which are the precious necessaries of life to her; or I turn to that village wedding, kept between four brown walls, where an awkward bridegroom opens the dance with a high-shouldered, broad-faced bride, while elderly and middle-aged friends look on, with very irregular noses and lips, and probably with quart-pots in their hands, but with an expression of unmistakable contentment and good - will. "Foh!" says my idealistic friend, "what vulgar details! What good is there in taking all these pains to give an exact likeness of old women and clowns? What a low phase of life! what clumsy, ugly people!"

But, bless us, things may be lovable that are not altogether handsome, I hope? I am not at all sure that the majority of the human race have not been ugly, and even among those "lords of their kind," the British, squat figures, ill-shapen nostrils, and dingy complexions are not startling exceptions. Yet there is a great deal of family love among us. I have a friend or two whose class of features is such that the Apollo curl on the summit of their brows would be decidedly trying; yet, to my certain knowledge, tender hearts have beaten for them, and their miniatures—flattering, but still not lovely—are kissed in secret by motherly lips. I have seen many an excellent matron who could never in her best days have been handsome, and yet she had a packet of yellow love-letters in a private drawer, and sweet children showered kisses on her sallow cheeks. And I believe there have been plenty of young heroes, of middle stature and feeble beards, who have felt quite sure they could never love anything more insignificant than a Diana, and yet have found themselves in middle life happily settled with a wife who waddles. Yes! thank God; human feeling is like the mighty rivers that bless the earth; it does not wait for beauty—it flows with resistless force, and brings beauty with it.

All honor and reverence to the divine beauty of form! Let us cultivate it to the utmost in men, women, and children—in our gardens and in our houses; but let us love that other beauty, too, which lies in no secret of proportion, but in the secret of deep human sympathy. Paint us an angel, if you can, with a floating violet robe, and a face paled by the celestial light; paint us yet oftener a Madonna, turning her mild face upward, and opening her arms to welcome the divine glory; but do not impose on us any æsthetic rules which shall banish from the region of Art those old women scraping carrots with their work-worn hands, those heavy clowns taking holiday in a dingy pot-house—those rounded backs and stupid,

35

BREVIER No. 12 SOLID
58 lines 739 words in page
12·7 words to line
Page 3½¾ × 6⅝ = 25·26 square inches
29·3 words to square inch
Page measures 61 × 35 = 2135 ems

S. W. GREEN'S SON
COMPOSITION, ELECTROTYPING, PRESSWORK
AND BINDING
74 and 76 Beekman Street

So I am content to tell my simple story, without trying to make things seem better than they were; dreading nothing, indeed, but falsity, which, in spite of one's best efforts, there is reason to dread. Falsehood is so easy, truth so difficult. The pencil is conscious of a delightful facility in drawing a griffin—the longer the claws, and the larger the wings, the better; but that marvellous facility, which we mistook for genius, is apt to forsake us when we want to draw a real unexaggerated lion. Examine your words well, and you will find that even when you have no motive to be false, it is a very hard thing to say the exact truth, even about your own immediate feelings—much harder than to say something fine about them which is *not* the exact truth.

It is for this rare, precious quality of truthfulness that I delight in many Dutch paintings, which lofty-minded people despise. I find a source of delicious sympathy in these faithful pictures of a monotonous homely existence, which has been the fate of so many more among my fellow-mortals than a life of pomp or of absolute indigence, of tragic suffering or of world-stirring actions. I turn without shrinking, from cloud-borne angels, from prophets, sibyls, and heroic warriors, to an old woman bending over her flower-pot, or eating her solitary dinner, while the noonday light, softened, perhaps, by a screen of leaves, falls on her mob-cap, and just touches the rim of her spinning-wheel, and her stone jug, and all those cheap, common things which are the precious necessaries of life to her; or I turn to that village wedding, kept between four brown walls, where an awkward bridegroom opens the dance with a high-shouldered, broad-faced bride, while elderly and middle-aged friends look on, with very irregular noses and lips, and probably with quart-pots in their hands, but with an expression of unmistakable contentment and good-will. "Foh!" says my idealistic friend, "what vulgar details! What good is there in taking all these pains to give an exact likeness of old women and clowns? What a low phase of life! what clumsy, ugly people!"

But, bless us, things may be lovable that are not altogether handsome, I hope? I am not at all sure that the majority of the human race have not been ugly, and even among those "lords of their kind," the British, squat figures, ill-shapen nostrils, and dingy complexions are not startling exceptions. Yet there is a great deal of family love among us. I have a friend or two whose class of features is such that the Apollo curl on the summit of their brows would be decidedly trying; yet, to my certain knowledge, tender hearts have beaten for them, and their miniatures—flattering, but still not lovely—are kissed in secret by motherly lips. I have seen many an excellent matron who could never in her best days have been handsome, and yet she had a packet of yellow love-letters in a private drawer, and sweet children showered kisses on her sallow cheeks. And I believe there have been plenty of young heroes, of middle stature and feeble beards, who have felt quite sure they could never love anything more insignificant than a Diana, and yet have found themselves in middle life happily settled with a wife who waddles. Yes! thank God; human feeling is like the mighty

36

BREVIER O. S. No. 3 LEADED
46 lines 584 words in page
12·7 words to line
Page 3⅔ × 6⅞ = 25·26 square inches
23·1 words to square inch
Page measures 61 × 35 = 2135 ems

S. W. GREEN'S SON
COMPOSITION, ELECTROTYPING, PRESSWORK
AND BINDING
74 and 76 Beekman Street

So I am content to tell my simple story, without trying to make things seem better than they were; dreading nothing, indeed, but falsity, which, in spite of one's best efforts, there is reason to dread. Falsehood is so easy, truth so difficult. The pencil is conscious of a delightful facility in drawing a griffin—the longer the claws, and the larger the wings, the better; but that marvellous facility, which we mistook for genius, is apt to forsake us when we want to draw a real unexaggerated lion. Examine your words well, and you will find that even when you have no motive to be false, it is a very hard thing to say the exact truth, even about your own immediate feelings—much harder than to say something fine about them which is *not* the exact truth.

It is for this rare, precious quality of truthfulness that I delight in many Dutch paintings, which lofty-minded people despise. I find a source of delicious sympathy in these faithful pictures of a monotonous homely existence, which has been the fate of so many more among my fellow-mortals than a life of pomp or of absolute indigence, of tragic suffering or of world-stirring actions. I turn without shrinking, from cloud-borne angels, from prophets, sibyls, and heroic warriors, to an old woman bending over her flower-pot, or eating her solitary dinner, while the noonday light, softened, perhaps, by a screen of leaves, falls on her mob-cap, and just touches the rim of her spinning-wheel, and her stone jug, and all those cheap, common things which are the precious necessaries of life to her; or I turn to that village wedding, kept between four brown walls, where an awkward bridegroom opens the dance with a high-shouldered, broad-faced bride, while elderly and middle-aged friends look on, with very irregular noses and lips, and probably with quart-pots in their hands, but with an expression of unmistakable contentment and good-will. "Foh!" says my idealistic friend, "what vulgar details! What good is there in taking all these pains to give an exact likeness of old women and clowns? What a low phase of life! what clumsy, ugly people!"

But, bless us, things may be lovable that are not altogether handsome, I hope? I am not at all sure that the majority of the human race have not been ugly, and even among those "lords of their kind," the British, squat figures, ill-shapen nostrils, and dingy complexions are not startling exceptions. Yet there is a great deal of family love among us. I have a friend or two whose class of features is such that the Apollo curl on the summit of their brows would be decidedly trying; yet, to my certain knowledge, tender hearts have beaten for them, and their miniatures—flattering, but still not lovely—are kissed in secret by motherly lips. I have seen many an excellent matron who could never in her best days have been handsome, and yet she had a packet of yellow love-letters in a private drawer, and sweet children showered kisses on her sallow cheeks. And I believe there have been plenty of young heroes, of middle stature and feeble beards, who have felt quite sure they could never love anything more insignificant than a Diana, and yet have found themselves in middle life happily settled with a wife who waddles. Yes! thank God; human feeling is like the mighty rivers that bless the earth; it does not wait for beauty—it flows with resistless force, and brings beauty with it.

All honor and reverence to the divine beauty of form! Let us cultivate it to the utmost in men, women, and children—in our gardens and in our houses; but let us love that other beauty, too, which lies in no secret of proportion, but in the secret of deep human sympathy. Paint us an angel, if you can, with a floating violet robe, and a face paled by the celestial light paint us yet oftener a Madonna, turning her mild face upward, and opening her arms to welcome the divine glory; but do not impose on us any æsthetic rules which shall banish from the region of Art those old women scraping carrots with their work-worn hands, those heavy clowns taking holiday in a dingy pot-house—those rounded backs and stupid, weather-beaten faces that

37

BREVIER O. S. No. 3 SOLID
58 lines 743 words in page
12·8 words to line
Page 3⅛ × 6⅚ = 25·26 square inches
29·5 words to square inch
Page measures 61 × 35 = 2135 ems
349 words to 1000 ems

S. W. GREEN'S SON
COMPOSITION, ELECTROTYPING, PRESSWORK
AND BINDING
74 and 76 Beekman Street
NEW YORK

MINION PLAIN FACE No. 20 LEADED.

So I am content to tell my simple story, without trying to make things seem better than they were; dreading nothing, indeed, but falsity, which, in spite of one's best efforts, there is reason to dread. Falsehood is so easy, truth so difficult. The pencil is conscious of a delightful facility in drawing a griffin—the longer the claws, and the larger the wings, the better; but that marvellous facility, which we mistook for genius, is apt to forsake us when we want to draw a real unexaggerated lion. Examine your words well, and you will find that even when you have no motive to be false, it is a very hard thing to say the exact truth, even about your own immediate feelings—much harder than to say something fine about them which is *not* the exact truth.

It is for this rare, precious quality of truthfulness that I delight in many Dutch paintings, which lofty-minded people despise. I find a source of delicious sympathy in these faithful pictures of a monotonous homely existence, which has been the fate of so many more among my fellow-mortals than a life of pomp or of absolute indigence, of tragic suffering or of world-stirring actions. I turn without shrinking, from cloud-borne angels, from prophets, sibyls, and heroic warriors, to an old woman bending over her flower-pot, or eating her solitary dinner, while the noonday light, softened, perhaps, by a screen of leaves, falls on her mob-cap, and just touches the rim of her spinning-wheel, and her stone jug, and all those cheap, common things which are the precious necessaries of life to her; or I turn to that village wedding, kept between four brown walls, where an awkward bridegroom opens the dance with a high-shouldered, broad-faced bride, while elderly and middle-aged friends look on, with very irregular noses and lips, and probably with quart-pots in their hands, but with an expression of unmistakable contentment and good-will. "Foh!" says my idealistic friend, "what vulgar details! What good is there in taking all these pains to give an exact likeness of old women and clowns? What a low phase of life! what clumsy, ugly people!"

But, bless us, things may be lovable that are not altogether handsome, I hope? I am not at all sure that the majority of the human race have not been ugly, and even among those "lords of their kind," the British, squat figures, ill-shapen nostrils, and dingy complexions are not startling exceptions. Yet there is a great deal of family love among us. I have a friend or two whose class of features is such that the Apollo curl on the summit of their brows would be decidedly trying; yet, to my certain knowledge, tender hearts have beaten for them, and their miniatures—flattering, but still not lovely—are kissed in secret by motherly lips. I have seen many an excellent matron who could never in her best days have been handsome, and yet she had a packet of yellow love-letters in a private drawer, and sweet children showered kisses on her sallow cheeks. And I believe there have been plenty of young heroes, of middle stature and feeble beards, who have felt quite sure they could never love anything more insignificant than a Diana, and yet have found themselves in middle life happily settled with a wife who waddles. Yes! thank God; human feeling is like the mighty rivers that bless the earth; it does not wait for beauty—it flows with resistless force, and brings beauty with it.

All honor and reverence to the divine beauty of form! Let us cultivate it to the utmost in men, women, and children—in our gardens and in our houses; but let us love that other beauty, too, which lies in no secret of proportion, but in the secret of deep human sympathy. Paint us an angel, if you can, with a floating violet robe, and a face paled by the celestial light; paint us yet oftener a Madonna, turning her mild face upward, and opening her arms to welcome the

38

MINION No. 20 LEADED
51 lines 695 words in page
13·6 words to line
Page 3⅛ × 6⅝ = 25·26 square inches
27·5 words to square inch
Page measures 69 × 40 = 2760 ems
252 words to 1000 ems

S. W. GREEN'S SON
COMPOSITION, ELECTROTYPING, PRESSWORK
AND BINDING
74 and 76 Beekman Street
NEW YORK

So I am content to tell my simple story, without trying to make things seem better than they were; dreading nothing, indeed, but falsity, which, in spite of one's best efforts, there is reason to dread. Falsehood is so easy, truth so difficult. The pencil is conscious of a delightful facility in drawing a griffin—the longer the claws, and the larger the wings, the better; but that marvellous facility, which we mistook for genius, is apt to forsake us when we want to draw a real unexaggerated lion. Examine your words well, and you will find that even when you have no motive to be false, it is a very hard thing to say the exact truth, even about your own immediate feelings—much harder than to say something fine about them which is *not* the exact truth.

It is for this rare, precious quality of truthfulness that I delight in many Dutch paintings, which lofty-minded people despise. I find a source of delicious sympathy in these faithful pictures of a monotonous homely existence, which has been the fate of so many more among my fellow-mortals than a life of pomp or of absolute indigence, of tragic suffering or of world-stirring actions. I turn without shrinking, from cloud-borne angels, from prophets, sibyls, and heroic warriors, to an old woman bending over her flower-pot, or eating her solitary dinner, while the noonday light, softened, perhaps, by a screen of leaves, falls on her mob-cap, and just touches the rim of her spinning-wheel, and her stone jug, and all those cheap, common things which are the precious necessaries of life to her; or I turn to that village wedding, kept between four brown walls, where an awkward bridegroom opens the dance with a high-shouldered, broad-faced bride, while elderly and middle-aged friends look on, with very irregular noses and lips, and probably with quart-pots in their hands, but with an expression of unmistakable contentment and good-will. "Foh!" says my idealistic friend, "what vulgar details! What good is there in taking all these pains to give an exact likeness of old women and clowns? What a low phase of life! what clumsy, ugly people!"

But, bless us, things may be lovable that are not altogether handsome, I hope? I am not at all sure that the majority of the human race have not been ugly, and even among those "lords of their kind," the British, squat figures, ill-shapen nostrils, and dingy complexions are not startling exceptions. Yet there is a great deal of family love among us. I have a friend or two whose class of features is such that the Apollo curl on the summit of their brows would be decidedly trying; yet, to my certain knowledge, tender hearts have beaten for them, and their miniatures—flattering, but still not lovely—are kissed in secret by motherly lips. I have seen many an excellent matron who could never in her best days have been handsome, and yet she had a packet of yellow love-letters in a private drawer, and sweet children showered kisses on her sallow cheeks. And I believe there have been plenty of young heroes, of middle stature and feeble beards, who have felt quite sure they could never love anything more insignificant than a Diana, and yet have found themselves in middle life happily settled with a wife who waddles. Yes! thank God; human feeling is like the mighty rivers that bless the earth; it does not wait for beauty—it flows with resistless force, and brings beauty with it.

All honor and reverence to the divine beauty of form! Let us cultivate it to the utmost in men, women, and children—in our gardens and in our houses; but let us love that other beauty, too, which lies in no secret of proportion, but in the secret of deep human sympathy. Paint us an angel, if you can, with a floating violet robe, and a face paled by the celestial light; paint us yet oftener a Madonna, turning her mild face upward, and opening her arms to welcome the' divine glory; but do not impose on us any æsthetic rules which shall banish from the region of Art those old women scraping carrots with their work-worn hands, those heavy clowns taking holiday in a dingy pot-house—those rounded backs and stupid, weather-beaten faces that have bent over the spade and done the rough work of the world—those homes with their tin pans, their brown pitchers, their rough curs, and their clusters of onions. In this world there are so many of these common, coarse people, who have no picturesque sentimental wretchedness! It is so needful we should remember their existence, else we may happen to leave them quite out of our religion and philosophy, and frame lofty theories which only fit a world of extremes. Therefore let Art always remind us of them; therefore let us always have men ready to give the loving pains of a life to the faithful representing of commonplace things—men who see beauty in these commonplace things, and delight in showing how kindly the light of heaven falls on them. There are few prophets in the world—few sublimely beautiful women—few heroes. I can't afford to give all my love and reverence

MINION No. 20 SOLID
66 lines 900 words in page
13·6 words to line
Page 3½½ × 6⅝ = 25·26 square inches
35·6 words to square inch
Page measures 69 × 40 = 2760 ems

S. W. GREEN'S SON
COMPOSITION, ELECTROTYPING, PRESSWORK
AND BINDING
74 and 76 Beekman Street

So I am content to tell my simple story, without trying to make things seem better than they were; dreading nothing, indeed, but falsity, which, in spite of one's best efforts, there is reason to dread. Falsehood is so easy, truth so difficult. The pencil is conscious of a delightful facility in drawing a griffin—the longer the claws, and the larger the wings, the better; but that marvellous facility, which we mistook for genius, is apt to forsake us when we want to draw a real unexaggerated lion. Examine your words well, and you will find that even when you have no motive to be false, it is a very hard thing to say the exact truth, even about your own immediate feelings—much harder than to say something fine about them which is *not* the exact truth.

It is for this rare, precious quality of truthfulness that I delight in many Dutch paintings, which lofty-minded people despise. I find a source of delicious sympathy in these faithful pictures of a monotonous homely existence, which has been the fate of so many more among my fellow-mortals than a life of pomp or of absolute indigence, of tragic suffering or of world-stirring actions. I turn without shrinking, from cloud-borne angels, from prophets, sibyls, and heroic warriors, to an old woman bending over her flower-pot, or eating her solitary dinner, while the noonday light, softened, perhaps, by a screen of leaves, falls on her mob-cap, and just touches the rim of her spinning-wheel, and her stone jug, and all those cheap, common things which are the precious necessaries of life to her; or I turn to that village wedding, kept between four brown walls, where an awkward bridegroom opens the dance with a high-shouldered, broad-faced bride, while elderly and middle-aged friends look on, with very irregular noses and lips, and probably with quart-pots in their hands, but with an expression of unmistakable contentment and good-will. "Foh!" says my idealistic friend, "what vulgar details! What good is there in taking all these pains to give an exact likeness of old women and clowns? What a low phase of life! what clumsy, ugly people!"

But, bless us, things may be lovable that are not altogether handsome, I hope? I am not at all sure that the majority of the human race have not been ugly, and even among those "lords of their kind," the British, squat figures, ill-shapen nostrils, and dingy complexions are not startling exceptions. Yet there is a great deal of family love among us. I have a friend or two whose class of features is such that the Apollo curl on the summit of their brows would be decidedly trying; yet, to my certain knowledge, tender hearts have beaten for them, and their miniatures—flattering, but still not lovely—are kissed in secret by motherly lips. I have seen many an excellent matron who could never in her best days have been handsome, and yet she had a packet of yellow love-letters in a private drawer, and sweet children showered kisses on her sallow cheeks. And I believe there have been plenty of young heroes, of middle stature and feeble beards, who have felt quite sure they could never love anything more insignificant than a Diana, and yet have found themselves in middle life happily settled with a wife who waddles. Yes! thank God; human feeling is like the mighty rivers that bless the earth; it does not wait for beauty—it flows with resistless force, and brings beauty with it.

All honor and reverence to the divine beauty of form! Let us cultivate it to the utmost in men, women, and children—in our gardens and in our houses; but let us love that other beauty, too, which lies in no secret of proportion, but in the secret of deep human sympathy. Paint us an angel, if you can, with a floating violet robe, and a face paled by the celestial light; paint us yet oftener a Madonna, turning her mild face upward, and opening her arms to welcome the divine glory; but do not impose on us any æsthetic rules which shall banish from the region of Art those old women scraping carrots with their work-worn hands, those heavy clowns taking holiday in a dingy pot-house—those rounded backs and stupid, weather-beaten faces that have bent over the spade and done the rough work of the world—those homes with their tin pans,

40

MINION O. S. No. 3 LEADED
51 lines 762 words in page
15 words to line
Page 3½½ × 6⅝ = 25·26 square inches
30·1 words to square inch
Page measures 69 × 40 = 2760 ems
276 words to 1000 ems

S. W. GREEN'S SON
COMPOSITION, ELECTROTYPING, PRESSWORK
AND BINDING
74 and 76 Beekman Street
NEW YORK

So I am content to tell my simple story, without trying to make things seem better than they were ; dreading nothing, indeed, but falsity, which, in spite of one's best efforts, there is reason to dread. Falsehood is so easy, truth so difficult. The pencil is conscious of a delightful facility in drawing a griffin—the longer the claws, and the larger the wings, the better ; but that marvellous facility, which we mistook for genius, is apt to forsake us when we want to draw a real unexaggerated lion. Examine your words well, and you will find that even when you have no motive to be false, it is a very hard thing to say the exact truth, even about your own immediate feelings—much harder than to say something fine about them which is *not* the exact truth.

It is for this rare, precious quality of truthfulness that I delight in many Dutch paintings, which lofty-minded people despise. I find a source of delicious sympathy in these faithful pictures of a monotonous homely existence, which has been the fate of so many more among my fellow-mortals than a life of pomp or of absolute indigence, of tragic suffering or of world-stirring actions. I turn without shrinking, from cloud-borne angels, from prophets, sibyls, and heroic warriors, to an old woman bending over her flower-pot, or eating her solitary dinner, while the noonday light, softened, perhaps, by a screen of leaves, falls on her mob-cap, and just touches the rim of her spinning-wheel, and her stone jug, and all those cheap, common things which are the precious necessaries of life to her ; or I turn to that village wedding, kept between four brown walls, where an awkward bridegroom opens the dance with a high-shouldered, broad-faced bride, while elderly and middle-aged friends look on, with very irregular noses and lips, and probably with quart-pots in their hands, but with an expression of unmistakable contentment and good-will. "Foh !" says my idealistic friend, "what vulgar details ! What good is there in taking all these pains to give an exact likeness of old women and clowns ? What a low phase of life ! what clumsy, ugly people !"

But, bless us, things may be lovable that are not altogether handsome, I hope ? I am not at all sure that the majority of the human race have not been ugly, and even among those "lords of their kind," the British, squat figures, ill-shapen nostrils, and dingy complexions are not startling exceptions. Yet there is a great deal of family love among us. I have a friend or two whose class of features is such that the Apollo curl on the summit of their brows would be decidedly trying ; yet, to my certain knowledge, tender hearts have beaten for them, and their miniatures—flattering, but still not lovely—are kissed in secret by motherly lips. I have seen many an excellent matron who could never in her best days have been handsome, and yet she had a packet of yellow love-letters in a private drawer, and sweet children showered kisses on her sallow cheeks. And I believe there have been plenty of young heroes, of middle stature and feeble beards, who have felt quite sure they could never love anything more insignificant than a Diana, and yet have found themselves in middle life happily settled with a wife who waddles. Yes ! thank God ; human feeling is like the mighty rivers that bless the earth ; it does not wait for beauty—it flows with resistless force, and brings beauty with it.

All honor and reverence to the divine beauty of form ! Let us cultivate it to the utmost in men, women, and children—in our gardens and in our houses ; but let us love that other beauty, too, which lies in no secret of proportion, but in the secret of deep human sympathy. Paint us an angel, if you can, with a floating violet robe, and a face paled by the celestial light ; paint us yet oftener a Madonna, turning her mild face upward, and opening her arms to welcome the divine glory ; but do not impose on us any æsthetic rules which shall banish from the region of Art those old women scraping carrots with their work-worn hands, those heavy clowns taking holiday in a dingy pot-house—those rounded backs and stupid, weather-beaten faces that have bent over the spade and done the rough work of the world—those homes with their tin pans, their brown pitchers, their rough curs, and their clusters of onions. In this world there are so many of these common, coarse people, who have no picturesque sentimental wretchedness ! It is so needful we should remember their existence, else we may happen to leave them quite out of our religion and philosophy, and frame lofty theories which only fit a world of extremes. Therefore let Art always remind us of them ; therefore let us always have men ready to give the loving pains of a life to the faithful representing of commonplace things—men who see beauty in these commonplace things, and delight in showing how kindly the light of heaven falls on them. There are few prophets in the world—few sublimely beautiful women—few heroes. I can't afford to give all my love and reverence to such rarities ; I want a great deal of those feelings for my everyday fellow-men, especially for the few in the foreground of the great multitude, whose faces I know, whose hands I touch, for whom I have to make way with kindly courtesy. Neither are picturesque lazzaroni or romantic criminals half so frequent as your common laborer, who gets his own bread, and eats it vulgarly but creditably with his own pocket-knife. It is more needful that I should have a fibre of

So I am content to tell my simple story, without trying to make things seem better than they were; dreading nothing, indeed, but falsity, which, in spite of one's best efforts, there is reason to dread. Falsehood is so easy, truth so difficult. The pencil is conscious of a delightful facility in drawing a griffin—the longer the claws, and the larger the wings, the better; but that marvellous facility, which we mistook for genius, is apt to forsake us when we want to draw a real unexaggerated lion. Examine your words well, and you will find that even when you have no motive to be false, it is a very hard thing to say the exact truth, even about your own immediate feelings—much harder than to say something fine about them which is *not* the exact truth.

It is for this rare, precious quality of truthfulness that I delight in many Dutch paintings, which lofty-minded people despise. I find a source of delicious sympathy in these faithful pictures of a monotonous homely existence, which has been the fate of so many more among my fellow-mortals than a life of pomp or of absolute indigence, of tragic suffering or of world-stirring actions. I turn without shrinking, from cloud-borne angels, from prophets, sibyls, and heroic warriors, to an old woman bending over her flower-pot, or eating her solitary dinner, while the noonday light, softened, perhaps, by a screen of leaves, falls on her mob-cap, and just touches the rim of her spinning-wheel, and her stone jug, and all those cheap, common things which are the precious necessaries of life to her; or I turn to that village wedding, kept between four brown walls, where an awkward bridegroom opens the dance with a high-shouldered, broad-faced bride, while elderly and middle-aged friends look on, with very irregular noses and lips, and probably with quart-pots in their hands, but with an expression of unmistakable contentment and good-will. "Foh!" says my idealistic friend, "what vulgar details! What good is there in taking all these pains to give an exact likeness of old women and clowns? What a low phase of life! what clumsy, ugly people!"

But, bless us, things may be lovable that are not altogether handsome, I hope? I am not at all sure that the majority of the human race have not been ugly, and even among those "lords of their kind," the British, squat figures, ill-shapen nostrils, and dingy complexions are not startling exceptions. Yet there is a great deal of family love among us. I have a friend or two whose class of features is such that the Apollo curl on the summit of their brows would be decidedly trying; yet, to my certain knowledge, tender hearts have beaten for them, and their miniatures—flattering, but still not lovely—are kissed in secret by motherly lips. I have seen many an excellent matron who could never in her best days have been handsome, and yet she had a packet of yellow love-letters in a private drawer, and sweet children showered kisses on her sallow cheeks. And I believe there have been plenty of young heroes, of middle stature and feeble beards, who have felt quite sure they could never love anything more insignificant than a Diana, and yet have found themselves in middle life happily settled with a wife who waddles. Yes! thank God; human feeling is like the mighty rivers that bless the earth; it does not wait for beauty—it flows with resistless force, and brings beauty with it.

All honor and reverence to the divine beauty of form! Let us cultivate it to the utmost in men, women, and children—in our gardens and in our houses; but let us love that other beauty, too, which lies in no secret of proportion, but in the secret of deep human sympathy. Paint us an angel, if you can, with a floating violet robe, and a face paled by the celestial light; paint us yet oftener a Madonna, turning her mild face upward, and opening her arms to welcome the divine glory; but do not impose on us any æsthetic rules which shall banish from the region of Art those old women scraping carrots with their work-worn hands, those heavy clowns taking holiday in a dingy pot-house—those rounded backs and stupid, weather-beaten faces that have bent over the spade and done the rough work of the world—those homes with their tin pans, their brown pitchers, their rough curs, and their clusters of onions. In this world there are so many of these common, coarse people, who have no picturesque sentimental wretchedness! It is so needful we should remember their existence, else we may happen to leave them quite out of our religion and philosophy, and frame lofty theories which only fit a world of extremes. Therefore let Art always remind us of them; therefore let us always have men ready to give the loving pains of a life to the faithful repre-

42

NONPAREIL No. 20 LEADED
58 lines 852 words in page
14·7 words to line
Page 3⅛ × 6¾ = 25·26 square inches
33·7 words to square inch
Page measures 80 × 46 = 3680 ems

S. W. GREEN'S SON
COMPOSITION, ELECTROTYPING, PRESSWORK
AND BINDING
74 and 76 Beekman Street

So I am content to tell my simple story, without trying to make things seem better than they were; dreading nothing, indeed, but falsity, which, in spite of one's best efforts, there is reason to dread. Falsehood is so easy, truth so difficult. The pencil is conscious of a delightful facility in drawing a griffin—the longer the claws, and the larger the wings, the better; but that marvellous facility, which we mistook for genius, is apt to forsake us when we want to draw a real unexaggerated lion. Examine your words well, and you will find that even when you have no motive to be false, it is a very hard thing to say the exact truth, even about your own immediate feelings—much harder than to say something fine about them which is *not* the exact truth.

It is for this rare, precious quality of truthfulness that I delight in many Dutch paintings, which lofty-minded people despise. I find a source of delicious sympathy in these faithful pictures of a monotonous homely existence, which has been the fate of so many more among my fellow-mortals than a life of pomp or of absolute indigence, of tragic suffering or of world-stirring actions. I turn without shrinking, from cloud-borne angels, from prophets, sibyls, and heroic warriors, to an old woman bending over her flower-pot, or eating her solitary dinner, while the noonday light, softened, perhaps, by a screen of leaves, falls on her mob-cap, and just touches the rim of her spinning-wheel, and her stone jug, and all those cheap, common things which are the precious necessaries of life to her; or I turn to that village wedding, kept between four brown walls, where an awkward bridegroom opens the dance with a high-shouldered, broad-faced bride, while elderly and middle-aged friends look on, with very irregular noses and lips, and probably with quart-pots in their hands, but with an expression of unmistakable contentment and good-will. "Foh!" says my idealistic friend, "what vulgar details! What good is there in taking all these pains to give an exact likeness of old women and clowns? What a low phase of life! what clumsy, ugly people!"

But, bless us, things may be lovable that are not altogether handsome, I hope? I am not at all sure that the majority of the human race have not been ugly, and even among those "lords of their kind," the British, squat figures, ill-shapen nostrils, and dingy complexions are not startling exceptions. Yet there is a great deal of family love among us. I have a friend or two whose class of features is such that the Apollo curl on the summit of their brows would be decidedly trying; yet, to my certain knowledge, tender hearts have beaten for them, and their miniatures—flattering, but still not lovely—are kissed in secret by motherly lips. I have seen many an excellent matron who could never in her best days have been handsome, and yet she had a packet of yellow love-letters in a private drawer, and sweet children showered kisses on her sallow cheeks. And I believe there have been plenty of young heroes, of middle stature and feeble beards, who have felt quite sure they could never love anything more insignificant than a Diana, and yet have found themselves in middle life happily settled with a wife who waddles. Yes! thank God; human feeling is like the mighty rivers that bless the earth; it does not wait for beauty—it flows with resistless force, and brings beauty with it.

All honor and reverence to the divine beauty of form! Let us cultivate it to the utmost in men, women, and children—in our gardens and in our houses; but let us love that other beauty, too, which lies in no secret of proportion, but in the secret of deep human sympathy. Paint us an angel, if you can, with a floating violet robe, and a face paled by the celestial light; paint us yet oftener a Madonna, turning her mild face upward, and opening her arms to welcome the divine glory; but do not impose on us any æsthetic rules which shall banish from the region of Art those old women scraping carrots with their work-worn hands, those heavy clowns taking holiday in a dingy pot-house—those rounded backs and stupid, weather-beaten faces that have bent over the spade and done the rough work of the world—those homes with their tin pans, their brown pitchers, their rough curs, and their clusters of onions. In this world there are so many of these common, coarse people, who have no picturesque sentimental wretchedness! It is so needful we should remember their existence, else we may happen to leave them quite out of our religion and philosophy, and frame lofty theories which only fit a world of extremes. Therefore let Art always remind us of them; therefore let us always have men ready to give the loving pains of a life to the faithful representing of commonplace things—men who see beauty in these commonplace things, and delight in showing how kindly the light of heaven falls on them. There are few prophets in the world—few sublimely beautiful women—few heroes. I can't afford to give all my love and reverence to such rarities; I want a great deal of those feelings for my everyday fellow-men, especially for the few in the foreground of the great multitude, whose faces I know, whose hands I touch, for whom I have to make way with kindly courtesy. Neither are picturesque lazzaroni or romantic criminals half so frequent as your common laborer, who gets his own bread, and eats it vulgarly but creditably with his own pocket-knife. It is more needful that I should have a fibre of sympathy connecting me with that vulgar citizen who weighs out my sugar in a vilely assorted cravat and waistcoat, than with the handsomest rascal in red scarf and green feathers; more needful that my heart should swell with loving admiration at some trait of gentle goodness in the faulty people who sit at the same hearth with me, or in the clergyman of my own parish, who is, perhaps, rather too corpulent, and in other respects is not an Oberlin or a Tillotson, than at the deeds of heroes whom I shall never know except by hearsay, or at the sublimest abstract of all clerical graces that was ever conceived by an able novelist.

And so I come back to Mr. Irwine, with whom I desire you to be in perfect charity, far as he may be from satisfying your demands on the clerical character. Perhaps

43

NONPAREIL No. 20 SOLID
77 lines, 1131 words in page
14·7 words to line
Page 3⅞ × 6⅚ = 25.26 square inches
45 words to square inch
Page measures 46 × 80 = 3680 ems

S. W. GREEN'S SON
COMPOSITION, ELECTROTYPING, PRESSWORK
AND BINDING

74 and 76 Beekman Street
NEW YORK

So I am content to tell my simple story, without trying to make things seem better than they were ; dreading nothing, indeed, but falsity, which, in spite of one's best efforts, there is reason to dread. Falsehood is so easy, truth so difficult. The pencil is conscious of a delightful facility in drawing a griffin—the longer the claws, and the larger the wings, the better; but that marvellous facility, which we mistook for genius, is apt to forsake us when we want to draw a real unexaggerated lion. Examine your words well, and you will find that even when you have no motive to be false, it is a very hard thing to say the exact truth, even about your own immediate feelings—much harder than to say something fine about them which is *not* the exact truth.

It is for this rare, precious quality of truthfulness that I delight in many Dutch paintings, which lofty-minded people despise. I find a source of delicious sympathy in these faithful pictures of a monotonous homely existence, which has been the fate of so many more among my fellow-mortals than a life of pomp or of absolute indigence, of tragic suffering or of world-stirring actions. I turn without shrinking, from cloud-borne angels, from prophets, sibyls, and heroic warriors, to an old woman bending over her flower-pot, or eating her solitary dinner, while the noonday light, softened, perhaps, by a screen of leaves, falls on her mob-cap, and just touches the rim of her spinning-wheel, and her stone jug, and all those cheap, common things which are the precious necessaries of life to her ; or I turn to that village wedding, kept between four brown walls, where an awkward bridegroom opens the dance with a high-shouldered, broad-faced bride, while elderly and middle-aged friends look on, with very irregular noses and lips, and probably with quart-pots in their hands, but with an expression of unmistakable contentment and good-will. "Foh !" says my idealistic friend, "what vulgar details ! What good is there in taking all these pains to give an exact likeness of old women and clowns? What a low phase of life ! what clumsy, ugly people !"

But, bless us, things may be lovable that are not altogether handsome, I hope? I am not at all sure that the majority of the human race have not been ugly, and even among those "lords of their kind," the British, squat figures, ill-shapen nostrils, and dingy complexions are not startling exceptions. Yet there is a great deal of family love among us. I have a friend or two whose class of features is such that the Apollo curl on the summit of their brows would be decidedly trying; yet, to my certain knowledge, tender hearts have beaten for them, and their miniatures—flattering, but still not lovely—are kissed in secret by motherly lips. I have seen many an excellent matron who could never in her best days have been handsome, and yet she had a packet of yellow love-letters in a private drawer, and sweet children showered kisses on her sallow cheeks. And I believe there have been plenty of young heroes, of middle stature and feeble beards, who have felt quite sure they could never love anything more insignificant than a Diana, and yet have found themselves in middle life happily settled with a wife who waddles. Yes ! thank God ; human feeling is like the mighty rivers that bless the earth ; it does not wait for beauty—it flows with resistless force, and brings beauty with it.

All honor and reverence to the divine beauty of form ! Let us cultivate it to the utmost in men, women, and children—in our gardens and in our houses ; but let us love that other beauty, too, which lies in no secret of proportion, but in the secret of deep human sympathy. Paint us an angel, if you can, with a floating violet robe, and a face paled by the celestial light ; paint us yet oftener a Madonna, turning her mild face upward, and opening her arms to welcome the divine glory ; but do not impose on us any æsthetic rules which shall banish from the region of Art those old women scraping carrots with their work-worn hands, those heavy clowns taking holiday in a dingy pot-house—those rounded backs and stupid, weather-beaten faces that have bent over the spade and done the rough work of the world—those homes with their tin pans, their brown pitchers, their rough curs, and their clusters of onions. In this world there are so many of these common, coarse people, who have no picturesque sentimental wretchedness ! It is so needful we should remember their existence, else we may happen to leave them quite out of our religion and philosophy, and frame lofty theories which only fit a world of extremes. Therefore let Art always remind us of them; therefore let us always have men ready to give the loving pains of a life to the faithful representing of commonplace things—men who see beauty in these commonplace things, and delight in showing how kindly the light of heaven falls on them. There are few prophets in the world—few sublimely beautiful women—

44

NONPAREIL O. S No. 3 LEADED
58 lines 887 words in page
15·3 words to line
Page 3⅛ × 6⅝ = 25·26 square inches
35·1 words to square inch
Page measures 80 × 46 = 3680 ems
24·1 words to 1000 ems

S. W. GREEN'S SON
COMPOSITION, ELECTROTYPING, PRESSWORK
AND BINDING
74 and 76 Beekman Street
NEW YORK

So I am content to tell my simple story, without trying to make things seem better than they were ; dreading nothing, indeed, but falsity, which, in spite of one's best efforts, there is reason to dread. Falsehood is so easy, truth so difficult. The pencil is conscious of a delightful facility in drawing a griffin—the longer the claws, and the larger the wings, the better; but that marvellous facility which we mistook for genius, is apt to forsake us when we want to draw a real unexaggerated lion. Examine your words well, and you will find that even when you have no motive to be false, it is a very hard thing to say the exact truth, even about your own immediate feelings—much harder than to say something fine about them which is *not* the exact truth.

It is for this rare, precious quality of truthfulness that I delight in many Dutch paintings, which lofty-minded people despise. I find a source of delicious sympathy in these faithful pictures of a monotonous homely existence, which has been the fate of so many more among my fellow-mortals than a life of pomp or of absolute indigence, of tragic suffering or of world-stirring actions. I turn without shrinking, from cloud-borne angels, from prophets, sibyls, and heroic warriors, to an old woman bending over her flower-pot, or eating her solitary dinner, while the noonday light, softened, perhaps, by a screen of leaves, falls on her mob-cap, and just touches the rim of her spinning-wheel, and her stone jug, and all those cheap, common things which are the precious necessaries of life to her; or I turn to that village wedding, kept between four walls, where an awkward bridegroom opens the dance with a high-shouldered, broad-faced bride, while elderly and middle-aged friends look on, with very irregular noses and lips, and probably with quart-pots in their hands, but with an expression of unmistakable contentment and good-will. " Foh !" says my idealistic friend, " what vulgar details ! What good is there in taking all these pains to give an exact likeness of old women and clowns ? What a low phase of life ! what clumsy, ugly people !"

But, bless us, things may be lovable that are not altogether handsome, I hope ? I am not at all sure that the majority of the human race have not been ugly, and even among those " lords of their kind," the British, squat figures, ill-shapen nostrils, and dingy complexions are not startling exceptions. Yet there is a great deal of family love among us. I have a friend or two whose class of features is such that the Apollo curl on the summit of their brows would be decidedly trying; yet, to my certain knowledge, tender hearts have beaten for them, and their miniatures—flattering, but still not lovely—are kissed in secret by motherly lips. I have seen many an excellent matron who could never in her best days have been handsome, and yet she had a packet of yellow love-letters in a private drawer, and sweet children showered kisses on her sallow cheeks. And I believe there have been plenty of young heroes, of middle stature and feeble beards, who have felt quite sure they could never love anything more insignificant than a Diana, and yet have found themselves in middle life happily settled with a wife who waddles. Yes ! thank God ; human feeling is like the mighty rivers that bless the earth ; it does not wait for beauty—it flows with resistless force, and brings beauty with it.

All honor and reverence to the divine beauty of form ! Let us cultivate it to the utmost in men, women, and children—in our gardens and in our houses ; but let us love that other beauty, too, which lies in no secret of proportion, but in the secret of deep human sympathy. Paint us an angel, if you can, with a floating violet robe, and a face paled by the celestial light ; paint us yet oftener a Madonna, turning her mild face upward, and opening her arms to welcome the divine glory ; but do not impose on us any æsthetic rules which shall banish from the region of Art those old women scraping carrots with their work-worn hands, those heavy clowns taking holiday in a dingy pot-house—those rounded backs and stupid, weather-beaten faces that have bent over the spade and done the rough work of the world—those homes with their tin pans, their brown pitchers, their rough curs, and their clusters of onions. In this world there are so many of these common, coarse people, who have no picturesque sentimental wretchedness ! It is so needful we should remember their existence, else we may happen to leave them quite out of our religion and philosophy, and frame lofty theories which only fit a world of extremes. Therefore let Art always remind us of them; therefore let us always have men ready to give the loving pains of a life to the faithful representing of commonplace things—men who see beauty in these commonplace things, and delight in showing how kindly the light of heaven falls on them. There are few prophets in the world—few sublimely beautiful women—few heroes. I can't afford to give all my love and reverence to such rarities; I want a great deal of those feelings for my everyday fellow-men, especially for the few in the foreground of the great multitude, whose faces I know, whose hands I touch, for whom I have to make way with kindly courtesy. Neither are picturesque lazzaroni or romantic criminals half so frequent as your common laborer, who gets his own bread, and eats it vulgarly but creditably with his own pocket-knife. It is more needful that I should have a fibre of sympathy connecting me with that vulgar citizen who weighs out my sugar in a vilely assorted cravat and waistcoat, than with the handsomest rascal in red scarf and green feathers; more needful that my heart should swell with loving admiration at some trait of gentle goodness in the faulty people who sit at the same hearth with me, or in the clergyman of my own parish, who is, perhaps, rather too corpulent, and in other respects is not an Oberlin or a Tillotson, than at the deeds of heroes whom I shall never know except by hearsay, or at the sublimest abstract of all clerical graces that was ever conceived by an able novelist.

And so I come back to Mr. Irwine, with whom I desire you to be in perfect charity, far as he may be from satisfying your demands on the clerical character. Perhaps you think he was not—as he ought to have been—a living demonstration of the benefits attached to the national church? But I am not sure of that ; at least I know that the people in Broxton and Hayslope would have been very sorry to part with their clergyman, and that most

45

NONPAREIL O. S. No. 3 SOLID
77 lines, 1185 words in page
154 words in line
Page 3⅛ × 6¾ = 25.36 square inches
47 words to square inch
Page measures 46 × 80 = 3680 ems

S. W. GREEN'S SON
COMPOSITION, ELECTROTYPING, PRESSWORK
AND BINDING
74 and 76 Beekman Street

So I am content to tell my simple story, without trying to make things seem better than they were ; dreading nothing, indeed, but falsity, which, in spite of one's best efforts, there is reason to dread. Falsehood is so easy, truth so difficult. The pencil is conscious of a delightful facility in drawing a griffin—the longer the claws, and the larger the wings, the better ; but that marvellous facility, which we mistook for genius, is apt to forsake us when we want to draw a real unexaggerated lion. Examine your words well, and you will find that even when you have no motive to be false, it is a very hard thing to say the exact truth, even about your own immediate feelings—much harder than to say something fine about them which is not the exact truth.

It is for this rare, precious quality of truthfulness that I delight in many Dutch paintings, which lofty-minded people despise. I find a source of delicious sympathy in these faithful pictures of a monotonous homely existence, which has been the fate of so many more among my fellow-mortals than a life of pomp or of absolute indigence, of tragic suffering or of world-stirring actions. I turn without shrinking, from cloud-borne angels, from prophets, sibyls, and heroic warriors, to an old woman bending over her flower-pot, or eating her solitary dinner, while the noonday light, softened, perhaps, by a screen of leaves, falls on her mob-cap, and just touches the rim of her spinning-wheel, and her stone jug, and all those cheap, common things which are the precious necessaries of life to her ; or I turn to that village wedding, kept between four brown walls, where an awkward bridegroom opens the dance with a high-shouldered, broad-faced bride, while elderly and middle-aged friends look on, with very irregular noses and lips, and probably with quart-pots in their hands, but with an expression of unmistakable contentment and good-will. "Foh !" says my idealistic friend, "what vulgar details ! What good is there in taking all these pains to give an exact likeness of old women and clowns ? What a low phase of life ! what clumsy, ugly people !"

But, bless us, things may be lovable that are not altogether handsome, I hope ? I am not at all sure that the majority of the human race have not been ugly, and even among those "lords of their kind," the British, squat figures, ill-shapen nostrils, and dingy complexions are not startling exceptions. Yet there is a great deal of family love among us. I have a friend or two whose class of features is such that the Apollo curl on the summit of their brows would be decidedly trying ; yet, to my certain knowledge, tender hearts have beaten for them, and their miniatures—flattering, but still not lovely—are kissed in secret by motherly lips. I have seen many an excellent matron who could never in her best days have

been handsome, and yet she had a packet of yellow love-letters in a private drawer, and sweet children showered kisses on her sallow cheeks. And I believe there have been plenty of young heroes, of middle stature and feeble beards, who have felt quite sure they could never love anything more insignificant than a Diana, and yet have found themselves in middle life happily settled with a wife who waddles. Yes ! thank God ; human feeling is like the mighty rivers that bless the earth ; it does not wait for beauty—it flows with resistless force, and brings beauty with it.

All honor and reverence to the divine beauty of form ! Let us cultivate it to the utmost in men, women, and children — in our gardens and in our houses ; but let us love that other beauty, too, which lies in no secret of proportion, but in the secret of deep human sympathy. Paint us an angel, if you can, with a floating violet robe, and a face paled by the celestial light ; paint us yet oftener a Madonna, turning her mild face upward, and opening her arms to welcome the divine glory ; but do not impose on us any æsthetic rules which shall banish from the region of Art those old women scraping carrots with their work-worn hands, those heavy clowns taking holiday in a dingy pot-house—those rounded backs and stupid, weather-beaten faces that have bent over the spade and done the rough work of the world—those homes with their tin pans, their brown pitchers, their rough curs, and their clusters of onions. In this world there are so many of these common, coarse people, who have no picturesque sentimental wretchedness ! It is so needful we should remember their existence, else we may happen to leave them quite out of our religion and philosophy, and frame lofty theories which only fit a world of extremes. Therefore let Art always remind us of them ; therefore let us always have men ready to give the loving pains of a life to the faithful representing of commonplace things —men who see beauty in these commonplace things, and delight in showing how kindly the light of heaven falls on them. There are few prophets in the world—few sublimely beautiful women—few heroes. I can't afford to give all my love and reverence to such rarities ; I want a great deal of those feelings for my everyday fellow-men, especially for the few in the foreground of the great multitude, whose faces I know, whose hands I touch, for whom I have to make way with kindly courtesy. Neither are picturesque lazzaroni or romantic criminals half so frequent as your common laborer, who gets his own bread, and eats it vulgarly but creditably with his own pocket-knife. It is more needful that I should have a fibre of sympathy connecting me with that vulgar citizen who weighs out my sugar in a vilely assorted cravat and waistcoat, than with the handsomest rascal in red scarf and green feathers ; more needful that my heart should swell with loving admiration at some trait of gentle goodness in the faulty people who sit at the same hearth with me, or in the clergyman of my own parish, who is, perhaps, rather too corpulent, and in other respects is not an Oberlin or a Tillotson, than at the deeds of heroes whom I shall never know except by hearsay, or at the sublimest abstract of all clerical graces that was ever conceived by an able novelist.

And so I come back to Mr. Irwine, with whom I desire you to be in perfect charity, far as he may be from satisfying your demands on the clerical character. Perhaps you think he was not—as he ought to have been—a living demonstration of the benefits attached to the national church ? But I am not sure of that ; at least I know that the people in Broxton and Hayslope would have been very sorry to part with their clergyman, and that most faces brightened at his approach ; and until it can be proved that hatred is a better thing for the soul than love, I must believe that Mr. Irwine's influence in his parish was a more

AGATE No. 20 LEADED
64 lines, 504 words in column
7·8 words to line
Column 2 × 6½ = 13 square inches
39 words to square inch
Column measures 93 × 27 = 2511 ems
200 words to 1000 ems

AGATE No. 20 SOLID
89 lines, 723 words in column
8·1 words in line
Column 2 × 6½ = 13 square inches
55 words to square inch
Column measures 93 × 27 = 2511 en
288 words to 1000 ems

So I am content to tell my simple story, without trying to make things seem better than they were; dreading nothing, indeed, but falsity, which, in spite of one's best efforts, there is reason to dread. Falsehood is so easy, truth so difficult. The pencil is conscious of a delightful facility in drawing a griffin—the longer the claws, and the larger the wings, the better; but that marvellous facility, which we mistook for genius, is apt to forsake us when we want to draw a real unexaggerated lion. Examine your words well, and you will find that even when you have no motive to be false, it is a very hard thing to say the exact truth, even about your own immediate feelings—much harder than to say something fine about them which is not the exact truth.

It is for this rare, precious quality of truthfulness that I delight in many Dutch paintings, which lofty-minded people despise. I find a source of delicious sympathy in these faithful pictures of a monotonous homely existence, which has been the fate of so many more among my fellow-mortals than a life of pomp or of absolute indigence, of tragic suffering or of world-stirring actions. I turn without shrinking from cloud-borne angels, from prophets, sibyls, and heroic warriors, to an old woman bending over her flower-pot, or eating her solitary dinner, while the noonday light, softened, perhaps, by a screen of leaves, falls on her mob-cap, and just touches the rim of her spinning-wheel, and her stone jug, and all those cheap, common things which are the precious necessaries of life to her; or I turn to that village wedding, kept between four brown walls, where an awkward bridegroom opens the dance with a high-shouldered, broad-faced bride, while elderly and middle-aged friends look on, with very irregular noses and lips, and probably with quart-pots in their hands, but with an expression of unmistakable contentment and good-will. "Foh!" says my idealistic friend, "what vulgar details! What good is there in taking all these pains to give an exact likeness of old women and clowns! What a low phase of life! what clumsy, ugly people!"

But, bless us, things may be lovable that are not altogether handsome, I hope! I am not at all sure that the majority of the human race have not been ugly, and even among those "lords of their kind," the British, squat figures, ill-shapen nostrils, and dingy complexions are not startling exceptions. Yet there is a great deal of family love among us. I have a friend or two whose class of features is such that the Apollo curl on the summit of their brows would be decidedly trying; yet, to my certain knowledge, tender hearts have beaten for them, and their miniatures — flattering, but still not lovely—are kissed in secret by motherly lips. I have seen many an excellent matron who could never in her best days have been handsome, and yet she had a packet of yellow love-letters in a private drawer, and sweet children showered kisses on her sallow cheeks. And I believe there have been plenty of young heroes, of middle stature and feeble beards, who have felt quite sure they could never love anything more insignificant than a Diana, and yet have found themselves in middle life happily settled with a wife who waddles.

Yes! thank God; human feeling is like the mighty rivers that bless the earth; it does not wait for beauty—it flows with resistless force, and brings beauty with it.

All honor and reverence to the divine beauty of form! Let us cultivate it to the utmost in men, women, and children—in our gardens and in our houses; but let us love that other beauty, too, which lies in no secret of proportion, but in the secret of deep human sympathy. Paint us an angel, if you can, with a floating violet robe, and a face paled by the celestial light; paint us yet oftener a Madonna, turning her mild face upward, and opening her arms to welcome the divine glory; but do not impose on us any æsthetic rules which shall banish from the region of Art those old women scraping carrots with their work-worn hands, those heavy clowns taking holiday in a dingy pothouse—those rounded backs and stupid, weather-beaten faces that have bent over the spade and done the rough work of the world—those homes with their tin pans, their brown pitchers, their rough curs, and their clusters of onions. In this world there are so many of these common, coarse people, who have no picturesque sentimental wretchedness! It is so needful we should remember their existence, else we may happen to leave them quite out of our religion and philosophy, and frame lofty theories which only fit a world of extremes. Therefore let Art always remind us of them; therefore let us always have men ready to give the loving pains of a life to the faithful representing of commonplace things—men who see beauty in these commonplace things, and delight in showing how kindly the light of heaven falls on them. There are few prophets in the world—few sublimely beautiful women—few heroes. I can't afford to give all my love and reverence to such rarities; I want a great deal of those feelings for my everyday fellow-men, especially for the few in the foreground of the great multitude, whose faces I know, whose hands I touch, for whom I have to make way with kindly courtesy. Neither are picturesque lazzaroni or romantic criminals half so frequent as your common laborer, who gets his own bread, and eats it vulgarly but creditably with his own pocket-knife. It is more needful that I should have a fibre of sympathy connecting me with that vulgar citizen who weighs out my sugar in a vilely assorted cravat and waistcoat, than with the handsomest rascal in red scarf and green feathers; more needful that my heart should swell with loving admiration at some trait of gentle goodness in the faulty people who sit at the same hearth with me, or in the clergyman of my own parish, who is, perhaps, rather too corpulent, and in other respects is not an Oberlin or a Tillotson, than at the deeds of heroes whom I shall never know except by hearsay, or at the sublimest abstract of all clerical graces that was ever conceived by an able novelist.

And so I come back to Mr. Irwine, with whom I desire you to be in perfect charity, far as he may be from satisfying your demands on the clerical character. Perhaps you think he was not—as he ought to have been—A living demonstration of the benefits attached to the national church? But I am not sure of that; at least I know that the people in Broxton and Hayslope would have been very sorry to part with their clergyman, and that most faces brightened at his approach; and until it can be proved that hatred is a better thing for the soul than love, I must believe that Mr. Irwine's influence in his parish was a more wholesome one than that of the zealous Mr. Ryde, who came there twenty years afterward, when Mr. Irwine had been gathered to his fathers. It is true Mr. Ryde insisted strongly on the doctrines of the Reformation, visited his flock a great deal in their own homes, and was severe in rebuking the aberrations of the flesh—put a stop, indeed, to the Christmas rounds of the church singers, as promoting drunkenness and too light a handling of sacred things. But I gathered from Adam that, to whom I talked of these matters in his old age, that few clergymen could be less successful in winning the hearts of their parishioners than Mr. Ryde. They gathered a great many notions about doctrine from him, so that almost every church-goer under fifty began to distinguish as well between the genuine gospel and what did not come precisely up to that standard, as if he had been born and bred a Dissenter; and for some time after his arrival there seemed to be quite a religious movement in that quiet rural district. "But," said

47

PEARL No. 5 LEADED
67 lines, 575 words in column
8·6 words to line
Column 2 × 5½ = 13 square inches
44 words to square inch
olumn measures 29 × 98 = 2842 ems

PEARL No. 5 SOLID
95 lines, 828 words in column
8·7 words to line
Column 2 × 5½ = 13 square inches
64 words to square inch
Column measures 29 × 98 = 2842 ems

www.ingramcontent.com/pod-product-compliance
Lightning Source LLC
Chambersburg PA
CBHW022022190326
41519CB00010B/1573